8,10

The Quest for a Fusion Energy Reactor

The Quest for a Fusion Energy Reactor

An Insider's Account of the INTOR Workshop

Weston M. Stacey

OXFORD
UNIVERSITY PRESS
2010

OXFORD
UNIVERSITY PRESS

Oxford University Press, Inc., publishes works that further
Oxford University's objective of excellence
in research, scholarship, and education.

Oxford New York
Auckland Cape Town Dar es Salaam Hong Kong Karachi
Kuala Lumpur Madrid Melbourne Mexico City Nairobi
New Delhi Shanghai Taipei Toronto

With offices in
Argentina Austria Brazil Chile Czech Republic France Greece
Guatemala Hungary Italy Japan Poland Portugal Singapore
South Korea Switzerland Thailand Turkey Ukraine Vietnam

Copyright © 2010 by Oxford University Press, Inc.

Published by Oxford University Press, Inc.
198 Madison Avenue, New York, New York 10016
www.oup.com

Oxford is a registered trademark of Oxford University Press

Library of Congress Cataloging-in-Publication Data
Stacey, Weston M.
The quest for a fusion energy reactor : an insider's account of the INTOR
Workshop / Weston M. Stacey.
    p. cm.
Includes bibliographical references and index.
ISBN 978-0-19-973384-2
1. Fusion reactors—Design and construction.   2. Engineering test
reactors—Design and construction.   3. Tokamaks.   4. Fusion reactors—
Research—International coorporation.   5. International Tokamak Reactor
Workshop   I. Title.
TK9204.S62 2010
621.48'4—dc22   2009022620

9 8 7 6 5 4 3 2 1

Printed in the United States of America
on acid-free paper

To all of those who contributed
to the INTOR Workshop.

Acknowledgments

This book is in large part an account of scientific and technological information being collected, evaluated, and integrated into a design concept for a fusion reactor that was then analyzed in detail. Probably more than a thousand scientists and engineers in Europe, Japan, the USA, and the USSR were involved in this process, and the actual development of the underlying experimental data and theoretical concepts involved thousands of other scientists and engineers worldwide over a much longer period. The contributions of only a few hundred of these people who were the most active participants in the INTOR Workshop activities or leading the various government fusion programs during 1978–88 are recognized in this book, but without the work of the many other scientists and engineers who developed the basic information, the work of the INTOR Workshop could not have been carried out.

Several people have been instrumental in the production of the book. Phyllis Cohen, physics editor for Oxford University Press, had the insight to recognize the important story that was being told in a somewhat unconventional manner from reading a draft of the first chapters and has offered valuable advice on producing a final version of the book, particularly in choosing an informative title and by securing knowledgeable reviews of the manuscript with good suggestions for its improvement. Phyllis has also provided the essential guidance of the book through the production process. Trish Watson's copy editing was most helpful both in eliminating inconsistencies and improving syntax.

On the home front, Valarie Spradling has provided essential administrative support in producing electronic versions of drawings and photographs and in coordinating the transmission of the files involved in the production of this book. Finally, Drs. John Porter and Lucy Axtell provided comments on a draft of the first two chapters, which led to changes that make the material more accessible to the nonscientist reader.

Contents

The Quest for a Fusion Energy Reactor

I

Prologue (1978)

The multibillion dollar International Thermonuclear Experimental Reactor (ITER), for which construction began in 2009 following many years of research, development, design, and negotiation, is both a major step toward harnessing mankind's ultimate energy source, nuclear fusion, and an ambitious step toward bringing the nations of the world together to address a common challenge of our joint future—energy. The governments collaborating on ITER (the EU, Japan, Russia, the USA, Korea, China, India) represent more than half the population of the world.

The present ITER project has its origins in the INTOR Workshop (1978–88) in which fusion scientists and engineers from the European Community (EC), Japan, the USA, and the USSR joined together to assess the readiness of the world's fusion programs to undertake the design and construction of the first experimental fusion energy reactor, to define the research and development that would be necessary to do so, to develop a design concept for such a device, and to identify and analyze critical technical issues that would have to be overcome. It was on the basis of the positive results of the INTOR Workshop that Secretary Gorbachev made the recommendation to President Reagan at the 1985 Geneva summit that led to the formation of the ITER project.

In 1988 I wrote a scientific/technical summary of the INTOR Workshop (*Progress in Nuclear Energy,* vol. 11, p. 119, 1988). Now, twenty years later, perhaps enough time has passed to put into perspective the broader history of the INTOR Workshop and its role leading to the creation of the ITER project to build the first fusion energy reactor. This book is based on the working journal that I kept during the decade that I was the vice chairman of the INTOR Workshop, recording both the internal workings of the workshop and its external interactions with governmental bodies searching fitfully for the mechanisms of international cooperation. Some explanatory material is included to make both fusion and the history of the tortured path leading to the creation of a major international scientific project accessible to nonspecialists.

Energy Resources and the Rationale for Fusion Development

Nuclear fusion will almost surely become mankind's ultimate source of energy, because of the essentially limitless fuel source. One in every 10,000 water molecules contains an atom of the heavy form of hydrogen known as deuterium (D), so the oceanic fuel source for D+D fusion is essentially unlimited. However, fusion of D+D requires much higher temperatures to achieve the same fusion rate that can be achieved at lower (hence less difficult to achieve) temperatures by the fusion of deuterium with an even heavier form of hydrogen known at tritium (T). Since tritium is radioactive with a half-life of about 12 years, it does not exist in nature, but it can be made by neutron capture in the nucleus of lithium atoms. Because the products of the D+T fusion reaction are a helium nucleus and a neutron, the neutron produced by the fusion reaction can, in principle, be captured in lithium surrounding the fusion chamber to produce another T to replace the one destroyed in the fusion reaction, thus providing a self-sufficient fuel cycle for producing and using the tritium.

Because some of the neutrons produced by fusion will be captured in other materials or will leak from the system, and because some of the tritium will radioactively decay away before it can be used, it actually is necessary to have a few extra neutrons in order to produce enough tritium to make the D+T fusion fuel cycle self-sufficient. In this case, nature is beneficent in providing some materials (e.g., lead, beryllium) that, when they capture a neutron, emit two or three new neutrons. This neutron multiplication makes a self-sufficient D+T fusion fuel cycle possible. Thus, the ultimate, or limiting, fuel source for the D+T fusion reaction is lithium, and there is a lot of it. The best estimate that I know is that there is enough lithium to enable D+T fusion to provide all the electricity needed in the world for more than 6,000 years (at the estimated 2050 electricity usage rate). This seems to be a pretty good argument that the fuel source for fusion is "essentially unlimited."

The question of when fusion energy will be needed is much more complex. Most of the world's energy today is produced from carbon-based "fossil" fuels (coal, oil, gas, etc.). Even though the extent of these resources and the practicality and economics of their extraction (e.g., oil from tar sands) are still debated by "experts" and others, there are clearly limits on the remaining fossil fuel resources, and there is a substantial body of opinion that practical limits will be reached in the present century. It is also clear that there are adverse environmental effects both of extracting fossil fuels from the earth and of releasing carbon and sulfur into the atmosphere by burning them, so environmental limits on fossil fuels may be closer at hand than resource limits.

The most likely alternative to burning fossil fuel to produce energy, the nuclear fission of uranium, presently provides about 15% of the world's electricity, and there are strong indications that production will increase significantly in the coming decades. Again, there is uncertainty about the practical and economical limits of the extractable uranium (and thorium) resource, and there is a body of opinion that this fuel resource also will be exhausted this century if the current "once through" fuel cycle (which extracts only about 1% of the potential energy content of uranium) used worldwide (with a few exceptions) continues to be the norm. "Closing" the nuclear fuel cycle to extract much more (50–90%) of the potential energy content of uranium, by producing fissionable ^{239}PU by neutron transmutation of non-fissionable ^{238}U in special "breeder" reactors, could extend this fuel resource into the next century, but this possibility is not yet being implemented.

It is not clear that the "renewable" energy sources under discussion (solar, wind, biomass, etc.) can ever meet a significant fraction of the electricity need. Providing the projected electrical power needed for the USA alone in 2050 is estimated to require solar panels that cover about two-thirds the land area of the State of Georgia, or a few million very large wind turbines to be built, or the annual harvesting of a forest that covers more than the total land area of the USA.

More sophisticated analyses of this type have led the governments of the developed nations of the world to invest in nuclear fusion research over the past half century, joined more recently also by the developing nations. The major fusion programs of the world during most of this time were those in the USA, the USSR, Europe, and Japan, although smaller efforts existed in several other countries. More recently, South Korea, China, and India have significantly increased their efforts in fusion research.

Fusion research has now progressed to the point that conditions necessary for an energy-producing fusion reactor have been approached, and tens of thousands of kilowatts of thermal power have been produced by fusion experiments, albeit only for seconds. (A "fusion reactor" is basically an extension of these experiments to the integrated system of engineering components that is required to create and sustain the fusion reaction within a confined volume and to extract and convert to electricity the energy thereby produced.)

The International Thermonuclear Experimental Reactor

Construction of the first experimental fusion energy reactor, the International Thermonuclear Experimental Reactor (ITER), began in 2009 in France. ITER is a multibillion dollar project to build and operate internationally the penultimate fusion experiment on the path to fusion energy, an experiment that will achieve conditions in the plasma core (a gaseous mixture of deuterium and tritium ions and electrons at solar temperatures) that are sufficient to sustain in a small volume the thermonuclear processes that produce the energy of the sun and the stars. At the same time, ITER will test advanced engineering components that can be used in future fusion power reactors and demonstrate their ability to operate in such an extreme environment. After ITER, "demonstration" fusion energy reactors that produce electrical power on the grid in a dependable and practical fashion will be built next.

A joint project of the governments of the EU, Japan, the USA, Russia, China, India, and Korea, which represent altogether more than half the population of the earth, ITER is arguably the most significant effort at international scientific collaboration ever undertaken. In addition to a central team of hundreds of scientists and engineers assembled at the Cadarache construction site in the south of France, thousands of plasma physicists and other scientists and engineers of a multitude of disciplines, located in laboratories around the world, are involved in the ITER design verification, performance analyses, and construction and, more numerously, in the supporting research that will assure its success when it begins operation in 2018. Hundreds of industrial scientists and engineers in companies around the world are preparing to manufacture the various sophisticated components that will ultimately be assembled at Cadarache.

Fusion in the 1970s

The process leading to the construction of ITER began in the late 1970s at a time when local fusion programs in the USSR, the USA, Europe, and Japan were enjoying great success in achieving the required thermonuclear temperatures and in increasing the plasma pressure and the length of time that the energy within the plasma could be confined before escaping. The greatest progress was being made with plasmas confined in a toroidal (donut shape) magnetic configuration invented by the Russians and called a tokamak. A new generation of large tokamaks was under construction—the Tokamak Fusion Test Reactor (TFTR) in the USA, Tokamak-15 (T-15) in the USSR, Japan Tokamak 60 (JT60) in Japan, and the most powerful of them all, the Joint European Torus (JET) in the UK.

Already in the late 1970s, scientists and engineers at the Kurchatov Institute in Russia, at the Argonne and Oak Ridge National Laboratories and the General Atomics Company in the USA, and at the Japan Atomic Energy Research Institute (JAERI) in Nakamura were making exploratory designs of the tokamak experimental power reactors (EPRs) that would follow the coming generation of large tokamaks (TFTR, T-15, JET, JT60). I organized and led the design team at Argonne National Laboratory during the mid-1970s that produced two of the earliest EPR conceptual designs. Other EPR design teams in the USA at the same time were led by Mike Roberts at Oak Ridge National Laboratory and Charlie Baker at General Atomics.

Magnetic Confinement

The basic principle of magnetic confinement of the charged ions and electrons that make up a fusion plasma is straightforward, even if the more subtle implications are not. A magnetic field exerts a force (known as the Lorentz force) on a moving charged particle that is in a direction perpendicular to both the magnetic field direction and the particle direction of motion. This force causes the moving charged particle to change its direction of motion in such a way as to spiral about the magnetic field line with a radius that is inversely proportional to the strength of the magnetic field. In a fusion plasma this radius of spiral is a small fraction of an inch, so the charged particles essentially follow the magnetic field lines and spiral about them with a very small radius of spiral. Thus, the problem of "closed" magnetic confinement reduces to constructing a magnetic field configuration in which the field lines remain within the volume in which the plasma is to be confined and never intersect the wall surrounding that volume.

Electromagnetic fields can be produced by currents flowing in conductors (electromagnets). A rule of thumb for understanding electromagnetic fields is to make a fist with the right hand and then extend the thumb; an electrical current flowing in the direction of the extended thumb will produce an encircling magnetic field in the direction in which the fingers are curled. The simplest way to form a "closed" magnetic field is in a donut-shaped (toroidal) container with a conductor wrapped around it (like a child's Slinky toy bent around to close on itself). The current flowing in the conductor wrapped around the toroidal container will produce a "closed" magnetic field directed around the axis of the container and not intersecting with the wall. This simplest of magnetic confinement configurations is the basis of the tokamak.

The Tokamak

The tokamak configuration is illustrated in figure 1.1. A very hot gas of ions and electrons, known as a plasma, is confined by magnetic forces in a toroidal (donut-shaped) vacuum chamber. The plasma is heated to high temperatures either by the injection of beams of neutral particles that have been accelerated to high speeds (indicated in the figure) or by the energy of electromagnetic waves launched into the plasma.

The idea is that if enough ions can be confined at sufficiently high temperatures (high thermal speeds), then occasionally a deuterium and a tritium ion will approach each other with sufficient speeds to overcome the repulsive electrical force acting between these charged nuclei and come close enough together that the extremely strong, but extremely short-range, attractive nuclear force becomes dominant, causing the D and T ions to "fuse" together to form a "compound nucleus."

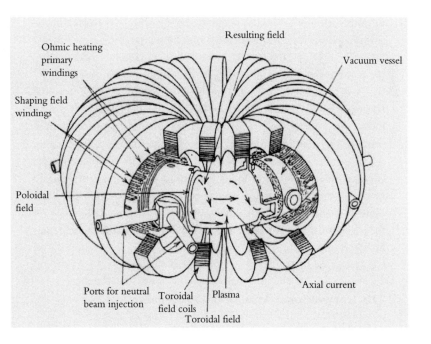

Figure 1.1 The tokamak configuration. (The "resulting field" is the sum of the toroidal and poloidal magnetic fields.)

This compound nucleus would be unstable and would immediately blow apart into a helium ion and a neutron, the combined masses of which are less than the combined masses of the deuterium and tritium ions that formed the compound nucleus. This excess mass would be converted to energy according to Einstein's famous equation $E = mc^2$.

This released "nuclear" energy would be in the form of kinetic energy (energy of motion) of the neutron (80%) and of the helium ion (20%). The helium ion, which is charged, is confined by the magnetic field force in the same way as the deuterium and tritium ions, with which it subsequently shares its energy via collisions. The neutron, on the other hand, is uncharged and thereby unaffected by either the magnetic field or the extremely low-density plasma medium, so it leaves the plasma chamber to interact with the surrounding material, sharing its energy via collisions with the atomic nuclei in those materials.

The plasma magnetic confinement in a tokamak is produced by a combination of a toroidal magnetic field circling the plasma in the toroidal (long way around) direction (shown by the arrow labeled "toroidal field" in figure 1.1) and a poloidal magnetic field circling the plasma in the poloidal (short way around) direction (indicated by the arrow labeled "poloidal field" in figure 1.1). The toroidal magnetic field is produced by a set of electromagnets called "toroidal field coils" encircling the plasma. The poloidal magnetic field is produced by a combination of an "axial" (toroidal) current, flowing around the plasma in the toroidal direction, and of other toroidal currents flowing in electromagnets outside the plasma (the "ohmic heating primary windings" and the "shaping field windings" indicated in figure 1.1). The resulting magnetic field, a combination of the toroidal and poloidal fields, spirals about the torus (like the stripes on a barber pole).

The International Atomic Energy Agency

International cooperation had been a characteristic of the world's fusion programs from the late 1950s, when the work on magnetic confinement of plasmas that had begun in World War II was declassified and presented at an International Atomic Energy Agency (IAEA) conference on the peaceful uses of nuclear energy. By the 1970s the IAEA, a UN agency with the primary mission of safeguarding nuclear materials, but with small scientific programs in several "nuclear" areas including fusion, was hosting a biennial conference and various specialist meetings and was publishing the research journal *Nuclear Fusion,* all of which were significant venues for international information exchange in fusion research.

The government officials with responsibility for the fusion development programs in those IAEA member countries that had them were members of the International Fusion Research Council (IFRC), a formal advisory body to the IAEA on its fusion activities, but in fact also a valuable informal venue for sharing information and working out small-scale cooperative arrangements. The USA was represented by Edwin Kintner, the USSR by Yevgeny Velikhov, and the Japanese representation changed from meeting to meeting to accommodate the dual university and government fusion programs in Japan but frequently included Sigeru Mori, the head of the JAERI fusion program. Europe, which was in a state of consolidation into the EC at the time, was represented by Donato Palumbo, the head of the EC fusion program, which consisted of several separate national programs (UK, Germany, France, Belgium, Holland, Sweden) whose heads also served on the IFRC. Australia was also represented. The leaders of the four major programs, USA, USSR, EC, and Japan (who jokingly referred to themselves as the "Gang of Four"), together with the chairman of the IFRC, served as an influential subcommittee of the larger IFRC.

In January 1978, the director general of the IAEA, Sigvard Ecklund of Sweden, invited member governments sponsoring fusion research to indicate their time scale for fusion development, their interest in increasing international cooperation, and their interest in participating in international studies of the next major step. Most of the responses were pro forma, thanking the IAEA for their excellent efforts and so forth, but the reply from Yevgeny Velikhov for the USSR was quite different. He proposed that the world's fusion programs join together under the auspices of the IAEA to jointly design, perform the supporting research and development, construct, and operate a first Experimental Power Reactor based on the tokamak concept.

The Soviet proposal was turned over to the IFRC. The reaction of the other IFRC members was guarded. Ed Kintner, then head of the U.S. Department of Energy (DoE) fusion program office, had in mind that the USA would build its own EPR, based on the exploratory design studies just completed in the USA, and was apprehensive that even talk of an international project could undermine the proposal that he was preparing, but he recognized the value of an international endorsement. The Japanese reaction was positive but cautious, at least in part because of the continuing dispute with the Soviet Union over the Kuril Islands. Donato Palumbo, the head of the EC fusion program office in Brussels, apparently viewed this proposal as a distraction to his efforts to pull together the separate national fusion programs in Europe and was initially opposed.

Fortunately for the future of ITER, the chairman of the IFRC was Rathbone Sebastian (Bas) Pease, an accomplished scientist and a talented galvanizer of committee action, then head of the U.K. fusion program. Taking advantage of the fact that the meeting was being held in his language, he masterfully synthesized these three and the equally diverse positions of the other IFRC members into a recommendation to ask the IAEA to form a "Specialist Committee" of international fusion experts to assess the technical readiness of the world's fusion programs to undertake the USSR proposal to construct and operate internationally this next major step in fusion development. The committee was to report its initial findings to the IFRC within one year. The authority for the organization and detailed guidance of the Specialist Committee and for the resolution of any issues upon which the specialists could not agree was delegated to a Steering Committee to consist of the leaders of the delegations from the EC, Japan, USSR, and USA.

The work of this Specialist Committee of fusion experts was to be performed in phases, and at the end of each phase the IFRC would determine whether to continue the Specialist Committee. At Palumbo's insistence, the first year was ignominiously designated the Zero Phase. The future ITER had cleared the first of many hurdles.

I first learned of this activity in the early fall of 1978 when Frank Coffman, a U.S. DoE fusion program manager with whom I had worked for several years while leading exploratory studies of the EPR at Argonne, began a phone conversation with the announcement that he was going to make me famous (which I recognized immediately as translating that he wanted me to do something for him). Frank went on to tell me that a group of fusion scientists from the USA, USSR, Japan, and EC were going to Trieste for six months to design a fusion reactor and that he wanted me to organize and lead the U.S. contingent. I realized that it did not make much sense to give people a job like this and then isolate them from their resources (computers, colleagues, experiments, reference libraries, etc.), but that's not something you tell your program manager (who administers your research funding). After a short conversation on the details, it seemed like something big and interesting was going on, so I agreed to take on the job, even though I had only recently moved to Georgia Tech to become a professor of nuclear engineering.

I maintained a working journal over the following decade of what became the INTOR Workshop. My original intention was to record the suggestions and positions on detailed physics, engineering, and organizational issues of the various participants in meetings, the conclusions and decisions taken on the issues under discussion, the action items arising from them, and so forth.

As the INTOR Workshop evolved, the scope of the journal became much broader and came to reflect a personal history of the INTOR Workshop: the technical and personal issues that dominated it, discussions among members and the accomplishments based upon them, conversations and arguments with international scientists and engineers to move the workshop forward, the tensions and stresses of a culturally diverse group of scientists and engineers learning to put aside their differences to become a team, the competition for resources to support the USA contribution to the workshop, the interactions with government fusion program leaders in IFRC meetings where the details of international cooperation in fusion were being painfully worked out by midlevel government officials with conflicting personal agendas—in short, the creation of what became the present ITER project. This book is based on my INTOR journal, with some explanatory material added to make the scientific and engineering aspects of the subject matter more accessible to nonspecialists, plus some personal anecdotes and reflections to provide a sense of the atmosphere in which these events were occurring.

USA, Fall 1978

My first task was to recruit a U.S. team. Frank Coffman and Ed Kintner of the U.S. DoE fusion program office passed the word to the U.S. fusion laboratory directors to help get the new activity started. It was obvious that a prominent plasma physicist needed to be involved, so my first call was to Mel Gottlieb, the director of the leading U.S. Plasma Physics Laboratory at Princeton. He had already talked with Paul Rutherford, head of the Princeton plasma theory group, and Paul had agreed to be part of the U.S. team. Jerry Kulcinski, a nuclear engineering professor at the University of Wisconsin and an expert in materials and fusion reactor conceptual design, was a natural choice for the materials and nuclear aspects of the work, and he was interested.

Frank Coffman suggested John Gilleland, who was just completing supervision of the construction of the DIII-D tokamak at General Atomics, to handle the engineering aspects of the work (magnets, heating systems, etc.). When I went to General Atomics to meet John, I first saw him as a hard hat and gray suit three stories below the observation deck for the DIII-D pit. When my guide pointed him out with, "That's Gilleland," adding under his breath, "He doesn't know about weekends," I decided that I had found the man for the job.

In preparation for the organizational meeting of the international committee scheduled for late November 1978, the members of the new U.S. team discussed how they might carry out such an activity to assess the readiness of the world fusion program to design, construct, and operate a tokamak experimental fusion energy reactor. The concept quickly evolved of teams of experts working with the resources available within the existing fusion institutions in their different countries to assess the status of the physics and technology development in various areas necessary for an EPR. We developed a preliminary structure for organizing the multitude of physics and technology areas involved in a tokamak EPR into about 18 topical areas. Each area included a set of topics within the same or related scientific and engineering disciplines. Each of these topics could be expected to fall within the purview of a single individual who could represent the results of the national assessment in an international forum. I also assembled as background material the major parameters from the EPR exploratory studies that had been performed in the USA to date.

The idea at which we arrived was that coordinated assessments would be carried out by the four "parties"—USA, USSR, Japan, and Europe (represented at that time by the EC)—each drawing upon the resources in their "national" fusion programs. The international coordination would be provided by a small number of representatives from each party (four was the number that had been suggested by the IFRC) who would meet every few months to compare the results of the national assessments carried out by the four "home teams" and to define specific "homework tasks" that would be performed over the next few months and then reviewed at the next meeting.

Thus, fusion science had developed to the point in the late 1970s that a major international assessment of the readiness of fusion to move forward to the building of a first experimental fusion energy reactor could be undertaken. It would prove to be an interesting experience.

2

Zero Phase of the INTOR Workshop (1978–80)

For the Zero Phase of the Specialist Committee, the International Fusion Research Council (IFRC) of the International Atomic Energy Agency (IAEA) provided "Terms of Reference." These directed the Specialist Committee "to draw on the capability in all countries to prepare a report to be submitted to the IFRC describing the technical objectives and nature of the next large fusion device of the tokamak type that could be constructed internationally."

In detail, the Committee should: 1) Review and discuss the results of existing studies of next-step proposals and experimental power reactors; 2) Survey the results of experiments, theory and associated technology planned to be available in the early 1980s; 3) Make recommendations of the aims, outline technical realization and resource requirements of a possible next step, indicating the alternatives considered; and 4) Identify the problem areas that need to be tackled before the construction of the next step.

These Terms of Reference went on to specify three programmatic objectives: "1) Take the maximum reasonable step beyond the present generation of experiments to demonstrate the scientific, technical and engineering feasibility of the generation of electricity by pure D-T fusion; 2) Include in a 'primitive sense' all systems and components for practical fusion power plants; 3) Provide test facilities for systems, components and materials for practical fusion power."

Abingdon, November 1978

Bas Pease, chairman of the IFRC, had invited me to stop by the Culham Laboratory in the UK on the way over to Vienna for the organizational meeting of the Specialist Committee, no doubt so that he could size me up. We discussed the U.S. team's concept of organizing the activity as a workshop that met periodically to define "homework tasks" that each team could carry out between meetings, using the full resources of their national fusion programs, in preparation for discussion at the next workshop meeting. He liked the idea and agreed with me that the IAEA headquarters in Vienna would be a more accessible spot to hold the workshop meetings than the IAEA's International School in Trieste that the IFRC had originally suggested. When these two suggestions were later presented to the IFRC, they were approved.

Vienna, November 1978

The organizational meeting of the four "national" leaders of the Specialist Committee was held November 20–23, 1978, in Vienna. At this time the IAEA headquarters in the old Grand Hotel on the Ringstrasse in Vienna was overflowing, and the organizational meeting was held in a small conference room in a building on the Boltzmangasse, which we all thought was quite appropriate given the prominence of Boltzman's equation in the theory of plasmas. The participants were Sigeru Mori, the leader of the Japan Atomic Energy Research Institute (JAERI) fusion program (who had previously represented Japan on the IFRC and had been appointed chairman of the Specialist Committee by his fellow IFRC members); Boris Kadomtsev, perhaps the leading tokamak theorist of the day and head of the principal USSR tokamak fusion program at the Kurchatov Institute of Atomic Energy in Moscow; Gunter Grieger, a plasma physicist and the head of the stellarator plasma confinement program at the Max-Planck-Institut für Plasmaphysik (IPP) near Munich; and myself, a plasma physicist and nuclear engineer from Georgia Tech who, having recently led the Argonne exploratory experimental power reactor (EPR) studies, had a good appreciation of the trade-offs and interactions among physics and technology that would be necessary to determine the physical characteristics of an EPR. Representing the IAEA was Jim Phillips, a plasma physicist from Los Alamos on assignment with the IAEA.

After introductory pleasantries and coffee, Mori started our meeting by announcing that "if Prof. Kadomtsev will provide us with the correct scaling law, we can go home and design the reactor." (Since the length of time that energy could be confined in the plasma could not then and can not now be predicted from first principles, it was standard practice to scale the measured "energy confinement time" among different tokamaks in terms of parameters such as plasma current and magnetic field, using empirical constants to make the scaling fit the measured results. There were a plethora of such scaling laws, and Mori was asking Kadomtsev to pick the "right" one.) Boris was as nonplussed as I was by this suggested mode of operation and demurred with the suggestion that this would better be discussed by a group of specialists.

In the momentary lull that followed, I brought up the U.S. concept of a series of periodic workshops in which this and other issues would be discussed by specialists from the four national groups, with work being done between meetings in the home country laboratories to provide input for these discussions. I distributed the preliminary workshop organizational structure that had been developed in the USA. This proposal struck a responsive chord, and a lively discussion ensued for the remainder of the day in exploration of the details of how this might work out. That evening we all dined at the nearby Hotel Atlanta, which I at least took to be a good omen.

We spent the next two days discussing details of the organization of the workshop into expert groups addressing different scientific and technical issues and identifying the probable performance objectives of this major next step in the world's fusion program. Sixteen physics and engineering topical groups, a Cost & Schedule group, and a Facilities & Personnel group were identified.

The identities of the four members of each "national" team were then discussed vis-à-vis the expertise that would be needed for representation in the (now) sixteen topical groups. Kadomtsev was pleased to learn that Paul Rutherford would represent the U.S. physics assessment at the workshop, and I suggested that Roger Hancox, the leading European fusion reactor conceptual designer, be added to the European Community (EC) team. Otherwise, the suggested names were accepted without comment. We also agreed to collect the results of the national exploratory studies for an EPR to serve as a guide for the assessment, and agreed that Vienna was a more practical site for the meetings than Trieste.

Since I had thought through beforehand the details that we were discussing on the workshop organization of the assessment, and since the meeting was being conducted in English, I naturally evolved as the de facto discussion leader. I believe that at this moment Mori found his solution to the problem of how he was going to run a workshop in a language in which he had some difficulty expressing himself. He proposed that I be the vice chairman of the workshop, and Grieger and Kadomtsev agreed.

Finally, we took up the momentous question of what acronym to adopt for our group. After a wide-ranging discussion, UNITOR was selected, since it conveyed, at least to us, that we were an International group working on a Tokamak Reactor under the auspices of the United Nations. Later, Jim Phillips (our IAEA scientific secretary) checked this out and informed us that any organization with UN in its name must be approved by the UN General Assembly. We decided not to go that route and settled on INTOR, which at least conveyed that we were an INternational group working on a TOkamak Reactor. Thus was born in a small conference room on the Boltzmangasse in Vienna in November 1978 the INTOR Workshop, which, working under the auspices of the IAEA of the UN, carried what eventually became the ITER project through its first decade.

The Steering Committee agreed that the members of the new INTOR Workshop should be prepared to discuss the scientific and technical issues and to define the principal questions to be examined in each of the sixteen categories at a first session in February and planned the agenda for that session. Then we all went home to pull together our initial assessment of the status of development in the sixteen identified areas of physics and engineering relative to what was needed to undertake the design and construction of a fusion EPR.

Physics and Technology Topical Areas

The sixteen topical areas around which the INTOR Workshop was initially organized were chosen to assess the status of the essential physics and engineering technologies that we perceived to underlie a power-producing tokamak experimental reactor.

The core of a fusion device is a very hot gas of electrically charged ions (of the heavier isotopes of hydrogen called deuterium and tritium) and electrons, in equal number so that the gas is macroscopically neutral in charge. This type of very hot gas of charged particles, known as a plasma, is the substance of the sun and stars.

As described in chapter 1, the fusion of two of these ions can only occur if they approach each other with a relative speed that is high enough to overcome the very large repulsive electrical force that acts between two positively charged ions (the nuclei of deuterium and tritium) and allow them to approach each other so closely that the very strong, but very short-range, attractive nuclear force can take over and "fuse" the two nuclei together. The "compound nucleus" so formed is energetically unstable and separates immediately into a neutron and a helium nucleus, the combined mass of which is less than the combined mass of the initial deuterium and tritium nuclei. The difference in mass is converted to energy, which can be recovered and converted into electricity.

In order for the thermal velocities of the deuterium and tritium nuclei to be large enough for them to overcome the repulsive electrical force between them and fuse, the plasma must be heated to "thermonuclear" temperatures (50–100 million degrees) similar to those found in the sun and stars. This had been accomplished in 1978 in the Princeton Large Torus (PLT) using neutral beam injectors in which ions were first accelerated electrically to high energies and then neutralized so that they could pass through the complex magnetic fields to enter the plasma, where they would again become energetic ions that transferred their energy to the plasma ions and electrons by colliding with them, thereby heating them.

However, an EPR would be larger and denser than the PLT, and the same neutral beams would not penetrate deeply enough to heat the EPR plasma. For beams at higher energy that would penetrate into the plasma of an EPR, the energy required to neutralize the ions, and thus the electrical power required to operate the neutral beam injectors, was prohibitively large and impractical.

Japan and the USA were developing a novel ion source to make negative ions (by electron attachment) for neutral beam injectors that could be neutralized with a much greater efficiency, thus requiring less power. In addition, there had been significant research on heating plasmas by launching electromagnetic waves at either radio frequency or microwave frequency into the plasma to heat it (in effect, making the plasma chamber a huge microwave oven). However, wave heating of plasmas had not yet achieved the plasma temperatures required for fusion, because of various physics and technological problems, although the USSR had apparently developed significantly higher power sources for generating microwaves than were available in the West. Thus, the assessment of the physics and technology of (1) *Plasma Heating* was an obvious high priority topical area for the INTOR Workshop.

The ions (nuclei of deuterium and tritium) in a plasma at a temperature sufficient for fusion will be moving with speeds of millions of meters per second. In a tokamak, magnetic fields are used to confine these rapidly moving ions (and electrons) in a toroidal container with dimensions of meters by means of the electromagnetic force that acts on a moving charged particle to change its direction of motion. Properly configured, this electromagnetic field can cause the ions and electrons to follow orbits that remain within the toroidal confinement chamber without colliding with the containment walls.

The electromagnets used to produce these confining forces in tokamaks (both at the time and, with few exceptions, today) were made with copper conductors. The resistive heating of these magnets in existing tokamaks was very large (in fact, magnet heating limited the time that plasma discharges could be maintained) and would be prohibitively large for an EPR, so it was clear that superconducting (zero-resistance) magnets would be required for an EPR. Exploratory design studies of fusion reactors had shown that magnetic field strengths of 10–11 Tesla led to much better designs than the magnetic field strengths of about 8 Tesla that could be achieved with the proven niobium-titanium superconductor that had been developed and used in magnets for high-energy physics accelerators and bubble chambers. The other known superconductor with which there was practical experience, niobium-tin, could achieve the higher fields, but it was brittle and unproven in large magnet applications. The technology of superconducting (2) *Magnets* was another obvious high priority topical area for the INTOR Workshop assessment.

Once the plasma is heated to fusion temperatures (50–100 million degrees), the fusion event itself will provide self-heating because the energetic helium nuclei produced in the fusion reactions are charged and thus are also magnetically confined within the toroidal plasma chamber, where they transfer their energy to the plasma ions and electrons by collisions. The plasma loses energy by radiation and by the transport of particles and energy out of the plasma onto the surrounding material walls.

In a practical, net power-producing fusion reactor, the high plasma temperatures will have to be maintained largely, if not entirely, by self-heating via fusion, with only a small amount of external neutral beam or electromagnetic wave heating. Demonstration that this was possible was an important task for an EPR. The amount of external heating required would depend on the magnitude of the radiation and transport cooling losses that must be compensated. The transport of particles and energy in tokamaks was then (and remains today) an area of active plasma physics research, and reliance on empirical scaling laws was necessary for prediction of how much external power would be needed for future machines. The radiation from a plasma depends very strongly on the amount of impurities in the plasma—ions with higher atomic numbers, such as iron or other material that "sputters" from the chamber wall from collisions with escaping plasma ions, that enter the plasma. Thus, (3) *Confinement* and (4) *Impurity Control* were both high-priority physics topics for the INTOR Workshop assessment.

The basic force balance on a tokamak plasma is between a confining magnetic pressure that would compress the plasma and an outward plasma gas kinetic pressure that would expand the plasma. Since the fusion power density (power per unit volume) increases with increasing plasma pressure, and cost increases with increasing magnetic field strength and size, a figure-of-merit for efficiency of plasma confinement is the ratio of the plasma pressure to the magnetic pressure, known as beta. A major thrust of tokamak plasma physics research at the time was to increase beta by finding ways to control incipient instabilities in the force balance equilibrium that would otherwise cause the plasma to lose confinement when beta rose above a certain value. Thus, achieving a plasma beta that projected to an economically attractive future fusion reactor was a generally accepted requirement for an EPR. For this reason, (5) *Stability Control* was identified as a physics topical area for assessment in the INTOR Workshop.

The startup, operation, and shutdown of a large EPR plasma with an internal fusion heating source was an area that had not at that time (nor much at the present time) been explored, and the large amounts of energy that must be transferred in and out of the electromagnet coils and energy storage systems were well beyond what had to date been dealt with in fusion. Groups on the physics of (6) *Startup, Burn & Shutdown* and on the technology of (7) *Energy Storage & Transfer* were formed to assess these topics.

The probable presence of wall-sputtered impurity atoms and the certain presence of an accumulating level of helium impurity atoms from the fusion reactions implied the necessity of continually exhausting some of the plasma from the chamber to remove these impurities and of continually fueling to replenish the plasma. The exhausted plasma would contain mostly the tritium and deuterium "fuel" for the fusion reaction, which must be recovered for reuse. A group on (8) *Fueling & Exhaust* was formed to assess the physics and technology of these processes.

Loss of tritium, which diffuses readily into hot metallic structures, was a concern because tritium availability was limited and because tritium is radioactive. A (9) *Tritium* topical group was formed to assess the availability of tritium and the technology for tritium recovery, processing, and storage.

The presence of high-energy fusion neutrons will introduce a new (for fusion) environment in which all components of an EPR have to operate. The performance of materials in a high-energy neutron flux, particularly in the first wall facing the plasma, and the shielding of sensitive components such as the magnets were new issues for fusion, albeit well-known issues in nuclear fission reactors. For this reason, topical groups were formed to assess (10) *Materials,* (11) *First-Wall,* and (12) *Shielding* technologies.

A topical group was formed to review (13) *Mechanical Design* requirements for an EPR, and a separate topical group was formed to assess technology for (14) *Remote Maintenance* of the geometrically complex tokamak configuration, which would be necessitated by the neutron activation of the structural material.

One of the principal missions envisioned for an EPR was to test various concepts for a lithium-containing tritium-breeding blanket. Tritium has a 12.5-year half-life for radioactive decay and will thus have to be produced in future fusion reactors by neutron capture in a lithium-containing material. Testing of such breeding blanket concepts was considered a high-priority mission for an EPR, and a (15) *Blanket* group was identified to assess the feasibility of doing so, as well as to evaluate the possibility of INTOR producing its own tritium supply.

In order to monitor the performance of the plasma and of the engineering systems, it was necessary to adapt standard diagnostic procedures to the high neutron flux, high temperature conditions expected in an EPR. A (16) *Diagnostics* group was identified for this purpose.

In addition to these technical topical groups, a Cost & Schedule group and a Facilities & Personnel group were identified as being necessary for the evaluation.

The Steering Committee assigned themselves the responsibility for developing a set of reference physical (size) and performance (e.g., magnetic field strength, plasma current) parameters for a major next-step device in order to guide the assessment.

Organizing the Assessments, Winter 1978–79

For the Soviets and the Japanese, who were represented in INTOR by the leaders of centrally managed fusion programs, the assessment of the readiness of the fusion program to undertake the design of an EPR was administratively a relatively straightforward (albeit technically challenging) matter of sending requests for evaluations of the sixteen areas of physics and technology down the chain of command and letting the existing systems set to work on these questions. There was, of course, the necessity of securing input from industry and from other research institutes, but the major parts of the assessments were carried out "in-house."

For the teams from the EC and the USA, whose Steering Committee members were people without high-level line management authority in their "national" fusion development systems (which in any case were far from centrally managed in fact if not in form), the administrative organization of such technical assessments was a vastly different matter.

In Europe, each of the various European fusion laboratories had chosen various fusion physics and engineering topics that they thought important to develop, so each laboratory had a vested interest in particular lines of research and assumed the prerogative to speak for it, at least within Europe. This would turn out to make it very difficult for the European delegates to INTOR to agree to any comparison of the relative status or promise of alternative technologies being developed in different laboratories (e.g., different technologies for heating the plasma or different technologies for breeding new tritium to replace that burned in the fusion process). Gunter Grieger had the difficult task of mediating within this framework to assemble a team and make an assessment that somehow respected these institutional prerogatives within each of the sixteen areas of physics and technology. He also had to work under the critical eye of Donato Palumbo, the head of the EC fusion program in Brussels, who was apprehensive about the entire activity.

In contrast to the other assessments, the American assessment was strictly a "bottom-up" affair. The four U.S. INTOR Participants (Paul Rutherford, Jerry Kulcinski, John Gilleland, and myself) met in mid-December at the U.S. Department of Energy (DoE) headquarters in Germantown, Maryland, with DoE program managers and Don Steiner of Oak Ridge, who was the leader of the recently formed U.S. team charged with further developing the EPR conceptual design for the U.S. "next-step" device that the DoE fusion program director Ed Kintner wanted to build. The U.S. EPR had been renamed the Engineering Test Facility (ETF), apparently to give it the appearance of a new initiative.

Kintner and his deputy, John Clarke, were supportive of this initial assessment activity, hoping to obtain, in effect, an international endorsement for the U.S. ETF project they were planning to propose to Congress. Leading U.S. experts on each of the sixteen INTOR topics were identified. The DoE program managers agreed to encourage those experts, whom they funded in the fusion program, to participate in the INTOR assessment.

The U.S. INTOR team met again at Georgia Tech in early January with Don Steiner and John Sheffield of Oak Ridge and with Paul Reardon, who was managing the construction of the largest U.S. tokamak, the Tokamak Fusion Test Reactor (TFTR) at the Princeton Plasma Physics Laboratory, and previously had managed the Isabelle accelerator at Brookhaven National Laboratory in Upton, New York. We identified the desired membership of the sixteen teams of experts who would perform the U.S. assessments in the sixteen topical areas that had been defined by the INTOR Workshop. This consisted altogether of about 100 leading U.S. fusion scientists and engineers. A coordinator was identified for each of the sixteen teams, designated as an "INTOR Consultant," who would work with one of the four U.S. INTOR participants and be responsible for organizing the U.S. assessments within his topical area. The initial consultants were Ron Parker and David Rose of MIT, Steiner and Sheffield of Oak Ridge, and Reardon of Princeton.

Each of the sixteen expert groups was requested to meet and prepare a written assessment of the status of the physics or technology in their topical areas and to identify the principal research and development (R&D) required to raise that status to the requisite level for the design and construction of a tokamak fusion EPR. This initial assessment was to be made relative to a set of reference parameters (dimensions, magnetic field strength, power output, heating power, etc.) that I had assembled from the recent U.S. EPR studies. Reports were compiled by each group, reviewed by the one of the four U.S. INTOR participants, and became the U.S. input to the first session of the INTOR Workshop.

Vienna, February 1979

On February 5, 1979, the sixteen "INTOR Workshop participants" to Session I of the Zero Phase of the INTOR Workshop convened in the IAEA headquarters in the old Grand Hotel on the Ringstrasse in Vienna. Participants in this first session were as follows (for abbreviations, see the glossary): EC—Gunter Grieger (IPP, Germany), Folker Englemann (FOM, Netherlands), Peter Reynolds (Culham, UK), Daniel Leger (CEA, France); Japan—Sigeru Mori, T. Hiraoka, K. Sako, and T. Tazima (JAERI); USSR—Boris Kadomtsev, Boris Kolbasov, Vladimir Pistunovich, Gely Shatalov (Kurchatov); and USA—Bill Stacey (Georgia Tech), John Gilleland (General Atomics), Gerry Kulcinski (Wisconsin), Paul Rutherford (Princeton). The EC team also included Roger Hancox (Culham, UK) as an expert and Robert Verbeek (EC Fusion Program Office, Brussels) as scientific secretary.

Because of the overcrowding of the IAEA headquarters, we were assigned office space in a nearby building on the Annagasse that had been built as an in-town palace by Prinz Eugen with the reward given him for successfully defending the city from the invading Turks in 1623. We occupied spacious rooms with large porcelain stoves from a bygone century and tall windows overlooking the gabled rooftops of central Vienna. figure 2.1 shows us gathered in our conference room.

A few steps from our doorway down the small cobblestoned Annagasse was the main pedestrian shopping street of Vienna, the Kärtnerstrasse. To the left about fifty yards away was the stately Staatsoper, and beyond, the Ringstrasse surrounding the central district of Vienna. To the right a hundred yards or so away was the central square of Vienna with the magnificent gothic cathedral of St. Stephan, the Stephansdom.

The material presented to this first session of the INTOR Workshop varied greatly from delegation to delegation, according to the different understandings of what it meant to assess the technical and scientific readiness to undertake the design and construction of a major EPR that would determine the future of fusion research in the world. First, there were different preconceptions of what an EPR actually should be. Participants from the USA, the USSR, and Japan,

Figure 2.1   Zero Phase INTOR Workshop Participants: Annagasse, Vienna, February 1979. Sitting left to right: Roger Hancox (UK), Boris Kolbasov (USSR), Gunter Grieger (FRG), Boris Kadomtsev (USSR), Sigeru Mori (Japan), Bill Stacey (USA), T. Hiraoka (Japan). Standing left to right: Henry Seligman (IAEA-UK), ? (IAEA-Japan), Jim Phillips (IAEA-USA), Dan Leger (France), Bob Verbeek (Brussels), Jerry Kulcinski (USA), Folker Engelmann (Netherlands), John Gilleland (USA), K. Sako (Japan), Paul Rutherford (USA), Vladimir Pistunovich (USSR), ? (IAEA-USSR), T. Tazima (Japan).

who had done relatively extensive exploratory design studies, had some definite ideas on the necessary performance parameters and the likely physical characteristics of an EPR, and these ideas turned out to be quite similar. However, with the exception of a small study in the UK, the EC had not done any significant work prior to INTOR in defining the characteristics of an EPR.

There was also a philosophical difference among the different teams on how to go about assessing the readiness of the world's fusion programs to design and build an EPR. The Japanese were strongly oriented toward just designing a reactor that would meet certain performance goals (if it worked) and then identifying the required

R&D to make it work, and their reports presented to Session I were full of design calculations in far greater detail than was found in the other national reports.

The orientation of the EC team was at the other extreme, toward assembling the physics and engineering performance parameters that had already been achieved and then identifying an experimental reactor based on a very cautious extrapolation of these parameters.

The USA and USSR orientations were intermediate, and similar. The experimental reactor performance parameters must be a significant but realistically achievable extrapolation of the existing database toward what was needed for a future fusion power reactor.

How much of an advance in physics or technology that was "realistically achievable" was a matter of judgment, which was colored by culture and individual experience. Different viewpoints began to emerge on the essential question of how much extrapolation beyond conditions already achieved would constitute an acceptable level of risk for the next major experiment in the world's fusion program. While this issue was of particular concern for the Europeans, who were more concerned than the rest of us that their colleagues back home would be watching them closely, waiting to criticize the slightest misjudgment, this was an issue over which the Participants continued to have legitimate differences over the duration of the workshop.

An important operational issue that arose during Session I was the authority of the different national INTOR Participants to agree to technical and scientific positions that evolved through group discussion at the INTOR Workshop but that differed from the original position presented to the workshop by the national team. The Japanese positions presented to the workshop had been evolved through the elaborate Japanese system of consensus building, and the Japanese Participants definitely did not feel authorized to formally agree with any different position that evolved from detailed discussions in the workshop, no matter their personal agreement in the matter. This situation persisted to some extent throughout the INTOR Workshop, with the Japanese unable to agree with a new position evolved at the workshop until it had worked through their system at home between sessions and then was presented as a Japanese recommendation at the next session, a phenomenon that came to be known as the "Japanese one session delay phenomenon."

The EC presentations to the workshop were crafted to respect the institutional interests of the various European fusion labs, and the EC INTOR Participants felt constrained to respect these institutional interests, to varying degrees, even when the other members of the workshop arrived at different positions that they privately supported. For the USA, almost no defending of institutional interests was involved, and a few phone calls to experts on the U.S. INTOR home team or to whomever we thought might know the most about the issue usually sufficed to confirm or nix any change in position purely on technical grounds. The Soviet Participants simply asked Kadomtsev's approval for any change in position with which they agreed technically.

The INTOR Steering Committee—Mori, Grieger, Kadomtsev, and myself—were nominally responsible for the organization and functioning of the workshop and had been empowered by the IFRC, hence by the IAEA, with the authority to act jointly as mediator or final arbitrator on any issues that could not be resolved by the normal decision-making procedures of the workshop—that is, by the agreement of the members of the responsible topical group. This proviso would turn out to be essential for moving the INTOR Workshop forward, but only if the Steering Committee members themselves were willing to exercise it when the need arose.

The Steering Committee met once every two to three days during Session I and throughout the following sessions of the workshop. The first two meetings were concerned with the functioning of the present session, but during the second week broader discussion of the coming year's work and of the accomplishment of the objectives of the workshop took precedence. These Steering Committee discussions were supplemented with one-on-one discussions among the members. These discussions were mainly exploratory, allowing the Steering Committee members to take a reading of each other.

It soon became clear that Kadomtsev and I had similar ideas about the objectives and probable characteristics of an EPR, shared similar outlooks on how the workshop should operate, agreed on the importance of a thorough assessment of the readiness of the physics and technology database to support an EPR design, and agreed on the important role INTOR could play in building support for an EPR in the international fusion community. Mori broadly agreed with our viewpoints, although he and the Japanese team were clearly impatient to get on with a design. We all three agreed on the importance of identifying with some specificity the R&D needed to make the necessary extensions of the physics and engineering database, of making priority rankings among alternative technologies, and of defining the EPR as a significant step beyond the present generation of large tokamaks.

Grieger was something of an enigma during these early Steering Committee meetings. He frequently objected on procedural grounds to procedural matters such as giving guidance to the topical groups intended to move their discussions forward, which he termed "interference" by the Steering Committee in the deliberations of the topical groups. We later realized that he was under pressure from some European laboratory directors not to have the INTOR Workshop arrive at unfavorable conclusions about the potential of research areas in which their laboratories were working, and he was under pressure from Palumbo, the EC fusion program leader, who wanted the workshop to end. However, we did not know this at the time, and some of his actions were inexplicable to us.

We all realized that we were dealing with issues that would ultimately determine the success or failure of the workshop. While we attempted to accommodate Grieger, the three of us who constituted a majority tacitly decided to move the workshop ahead according to our prevailing viewpoint.

However, we all understood not only that we had to make the workshop produce a sound and sensible result, but also that we needed to have all four parties on board. So, I invited Grieger to dinner, hoping that a more relaxed atmosphere would lead to better relations. We went to the Biereklinik, an old restaurant with Austrian specialties, on the Steindlgasse off the Graben in central Vienna. There was a Turkish cannon ball embedded in the whitewashed wall above our table, a relic of an earlier failed effort in Vienna.

After studying the menu, we agreed on what we would have, and Grieger offered to order. When the waiter arrived, he announced our order in his precise northern German, receiving in return a blank stare. The two of them exchanged a few words, then he told me with exasperation that the waiter was not even Austrian (which was bad enough in his view), but Romanian, and did not understand a word of German. "Well, neither do I, so let me try," I said and proceeded to order "zwei Weiner schnitzel mit Gemichter salat und zwei Bier, bitte" in my painful but practiced American Berlitz German. The waiter muttered something that neither of us understood and dashed off, returning eventually with just that. We both found this an amusing icebreaker and proceeded to enjoy a lingering dinner talking about everything except the INTOR Workshop. This was but the first of many such pleasant dinners the two of us shared over the course of the workshop.

*   *   *

The evolution in attitude from competitors into collaborators that the INTOR Workshop achieved among the participants began during Session I. At the outset of this first session, we certainly came together with the mindset of competitors. National pride, institutional self-interest, and plain old vainglory caused each Participant to favor technologies and physics concepts that were being developed in his country, perhaps even in his own laboratory, and to want these concepts to be chosen by the workshop. After two weeks of intense technical discussion, mitigated by informal discussions over coffee twice a day, by a couple of wine and cheese gatherings at the end of the day, and by dinners hosted by the Japanese and by the Soviets in candlelit Viennese restaurants, with frequently replenished pitchers of the good Austrian wine Grüner Veltliner, the spirit of international camaraderie and trust that was one of the greatest achievements of the INTOR Workshop began to take root.

*   *   *

In 1979, Austria had been geographically plunged into communist eastern Europe for quite some time, and we innocent Americans were anxious to see the other side of the iron curtain. Kulcinski and I took the train from Vienna across the empty, snowy plains to Budapest on our weekend in the February session, staying at a famous hot springs hotel built into the hillside of the Pest side of the Danube.

A Carnival celebration was in progress when we returned from dinner. We were barred from walking up the stairs from the lobby to the celebration, so we took the elevator to our floor and walked down a flight of stairs to join local communist dignitaries in a Catholic celebration that was still going strong when we left well after midnight. We got up a few hours later and went down to the famous baths, but our imaginations were dashed by the population of fat, bald male Hungarians puffing away on cigars.

*   *   *

As for the technical outcome of Session I, the accomplishments were the detailed identification of specific information that we agreed to compile in order (a) to identify a reference design concept for an EPR, (b) to make an assessment of the adequacy of the physics and technology database in the sixteen topical areas to support the design of such an EPR, and (c) to identify and prioritize the R&D needed in each of the sixteen topical areas in order to make the necessary extensions of the database. There was also agreement on a tentative set of "Guiding Parameters" for the INTOR concept that would be used to guide the assembly of this information. We all left Vienna with written documentation and a common understanding of this "homework" assignment for the following session. I think we also left with a realization that the enterprise we were embarked upon just might lead to something important.

USA, Spring 1979

Upon returning home in mid-February, the U.S. INTOR participants distributed the homework tasks to the U.S. INTOR home team and briefed them on the outcome of the first session and on the homework tasks that needed to be performed for the next session in mid-June. We expanded the membership and changed some of the leadership of the sixteen topical groups, both to reflect organizational changes that had emerged from the Vienna session and to engage those most interested in working on the tasks.

We put together a Concept Advisory Group consisting of most of the people in the USA who had done any tokamak reactor studies—Charlie Baker (formerly at General Atomics) from Argonne, Dan Cohn from MIT, Bob Conn from Wisconsin, Dan Jassby from Princeton, Martin Peng and Lowell Reid from Oak Ridge, John Rawls from General Atomics, and myself—to make calculations and develop recommendations for the likely parameters for an INTOR (EPR) concept.

In mid-March, I went to DoE headquarters in Germantown, Maryland, to brief Ed Kintner, John Clarke, Frank Coffman, and other DoE fusion program managers. Kintner, a career DoE bureaucrat with a technical background in nuclear engineering who had been Admiral Hyman Rickover's deputy in the highly successful Naval Reactors Program, was now serving as the head of a program in an area with which he was unfamiliar. He had recruited John Clarke from his position as head of the fusion program at Oak Ridge to be his deputy and heir apparent, with the promise of building an EPR, now incarnated in the recently renamed Engineering Test Facility (ETF) project located at Oak Ridge. Kintner informed me that he was pressing for ETF to be identified in the DoE fiscal year 1981 budget discussions, so it was important for him that the INTOR Workshop Zero Phase report be available in January 1980 to support those discussions. I informed him that this was the schedule and that I would try to make sure it happened.

Kintner was concerned about the pressure from the leading U.S. plasma physics laboratory at Princeton to build another long-pulse tokamak after their present TFTR experiment (which was approaching operational status) but before the ETF, which both Kintner and Clarke wanted to build next. Such a long-pulse experiment (without radioactive tritium and fusion neutrons) could be sited at Princeton, whereas the ETF would have to be sited at a larger and more remote site, such as Oak Ridge. (Since all aspects of plasma physics, except those involving the fusion event itself, can be investigated in deuterium plasmas without producing a high level of neutron activation of the facility, plasma physics experiments are normally carried out with pure deuterium plasmas, rather than with the combined deuterium-tritium plasmas that will be used in ITER and subsequent fusion reactors to produce fusion energy.)

As an expeditious alternative to the Princeton proposal, Kintner and Clarke were interested in possible upgrades of the present generation of large tokamaks soon to become operational (TFTR in the USA, T-15 in the USSR, JET in the EC, and JT60 in Japan). They wanted the INTOR Workshop to examine such upgrades for extending the physics database for ETF/INTOR. Kintner told me that there was no way that the U.S. fusion program could support a new long-pulse U.S. tokamak such as Princeton wanted or participation in an INTOR project at the same time that it was undertaking ETF. He made it clear where his priorities were, but it was far from clear that his plans would work out (they did not).

I left this meeting a bit perplexed about my professional responsibilities. I subsequently determined to do what I could to muster a strong U.S. participation in the INTOR Workshop despite Kintner's low priority for INTOR relative to a national U.S. ETF project. I rationalized that this course of action would keep the international option for an EPR alive and could not harm Kintner's aspirations for a domestic U.S. ETF. Needless to say, I did not repeat the gist of this conversation to my INTOR colleagues.

The 100+ fusion scientists and engineers from the major fusion laboratories, university programs, and industry that constituted the U.S. INTOR home team held several series of topical meetings throughout the spring of 1979, leading to the development of draft reports documenting the U.S. position on the homework tasks in the sixteen INTOR topical areas and on the INTOR design concept. These reports were presented to and debated by about fifty of the leading fusion people in the USA in these topical areas in a three-day meeting at Georgia Tech in late May, revised accordingly, and taken to Session II of the INTOR Workshop in June.

Vienna, June and July 1979

Session II of the INTOR Workshop was held in Vienna on June 11 through July 6, 1979. In the opening plenary session, Mori first reported that the IFRC (the IAEA advisory group to which the INTOR Workshop reported) had recently agreed that an IFRC subcommittee would meet with the INTOR Workshop at the following Session III, that the full IFRC would meet in January to review the INTOR Workshop report, and that the workshop should plan to be extended for the first four to six months of 1980 to make any necessary revisions to our report requested by the IFRC.

Then each national team summarized its findings and recommendations relevant to the homework tasks defined for the sixteen

INTOR topical areas and for the definition of the INTOR design concept. There was general support of the Guiding Parameters that had been identified in Session I to guide the evaluations, except for a recommendation for a somewhat larger physical size by the EC and recommendations by all parties that a "divertor" be incorporated to control unwanted "impurity" atoms that would be sputtered from the wall and radiatively cool the plasma.

In contrast to Session I, all of the delegations brought to Session II detailed analyses and data pertaining to answering the same set of questions and the evaluation of the same issues. At least the "homework tasks" aspect of the workshop procedure seemed to be working.

There was a definite advance relative to Session I on the clerical support front. During the previous session, all of the typing had been done by a single IAEA typist on an old electric typewriter. I decided that this was one of the old-world charms of Vienna that we could dispense with and made arrangements to have an IBM word processor delivered to us in Vienna prior to our arrival. All of the U.S. reports were put on floppy disks and taken to Vienna by my administrative assistant, Janie Griffith, who helped support Session II of the INTOR Workshop. The impending arrival of our word processor prompted the IAEA to also order one, so the clerical support for Session II was first-rate, and this greatly improved the productivity.

For certain areas in which major issues were anticipated to arise, additional experts had been asked to join the workshop for detailed discussions. Gorgi Churakov (Efremov Research Institute, USSR) and John Alcorn (General Atomics, USA) joined the workshop for discussion of superconducting magnets, and Dale Meade (Princeton, USA) and J. Bohdansky (IPP, Germany) joined the workshop for discussions of the divertor. Roger Hancox (Culham, UK) joined all sessions of the Zero Phase workshop for discussion of reactor design issues.

The first few weeks of Session II went quite well, with the members of the sixteen topical groups reading each other's reports and jointly developing detailed lists depicting the status of the various areas of physics and of the various engineering technologies involved. These groups attempted to formulate a common view on the status of development and required R&D in each topical area during the session. For example, the Plasma Heating group led by Rutherford compiled the recent experimental results on neutral beam heating technology, the physics results achieved on the heating of plasmas from several tokamaks (PLT, ORMAK, and ISX-B [USA]; DITE and TFR [EC]; and T-11 [USSR]), and the expected results from planned experiments on these devices and others that would become operational within the next few years. These results were compared against the anticipated needs of INTOR. The same exercise was then carried out for electromagnetic wave heating technologies, plasma heating experiments, and so forth, for all of the sixteen topical groups. This sort of comparison is frequently done in scientific conferences, albeit within a narrower scope, so the accomplishment of the INTOR Workshop to this point, while based on a much broader and more detailed assessment than had been heretofore performed and representing a worldwide effort, was not different in kind from what had come before.

The new and harder part began a couple of weeks into Session II. The INTOR Workshop was charged with evaluating the adequacy of this existing database, plus anticipated extensions, to undertake the design and construction of an EPR, and then to iden-tify the additional R&D needed. Most national fusion programs at this time were based on a "two steps to commercialization" strategy, with an EPR followed by a demonstration reactor, or DEMO, that in turn would be followed by commercial fusion power plants. This charge thus required that the workshop also identify the most prob-able characteristics and operating and performance parameters of an INTOR device that on the one hand could be based on plausible advances made within the next few years to the existing scientific and engineering database. On the other hand, this INTOR device would represent a significant enough advance in performance toward the DEMO and beyond the current generation of large toka-maks just beginning operation (e.g., TFTR, JET, JT60, T-15) to justify its being built at this time. This EPR (the INTOR) also needed to be consistent with the "two steps to commercialization" strategy, which required some consideration also of the second, DEMO step.

A meaningful response to such a task thus required making informed judgments about development priorities among different technologies for the various engineering systems and among different physics operating regimes for the plasma. Such judgments, by their nature, had to be made without full knowledge of the underlying physics and technology. This was a novel situation for members of the INTOR Workshop that engendered many lively discussions.

Setting priorities among alternative technologies or physics operating strategies was fraught with difficulties of a different type. Although the fusion programs of each of the four parties functioned within different domestic environments, in all the programs different technologies tended to be developed by different laboratories or industries, and an INTOR Workshop choice of one technology over another would inevitably have future repercussions in domestic resource prioritization, particularly in the less centralized fusion programs of the EC and USA.

Using again the example of plasma heating systems discussed above, a choice needed to be made among neutral beam injection heating, the several modes of electromagnetic wave heating, and a few other possible plasma heating technologies (e.g., transit time magnetic pumping and adiabatic compression, which had been demonstrated to heat plasmas, and relativistic electron beams, which had hardly been tried yet). Each of these technologies was being developed as a major program at one or more fusion laboratories, whose future prospects might be affected by how these technologies were rated by the INTOR Workshop.

The usual way to deal with such inconvenient difficulties when they arose at scientific conferences was just not to make priority choices, but instead to merely list all the alternative technologies or physics operating strategies as "options," and this solution began to surface in some of the topical area groups. I became increasingly concerned that if we failed to take priority decisions of this sort, we would forfeit any authority we might have in any recommendation that we might subsequently make to move ahead to a design project. In private discussions, I found that Kadomtsev and Mori agreed, but Grieger (who had to mediate among the interests of several European laboratories) was usually opposed to pressuring the topical groups to make priority choices.

The Steering Committee was empowered to resolve issues that could not be resolved by the members of the workshop, and at my instigation we agreed by majority vote to interpret this inability to make priority choices among alternatives as an issue on which the Steering Committee would intervene in the deliberations of the sixteen topical area groups. The Steering Committee members began attending group meetings in order to mediate the decision-making process, and in a few instances when this was not successful, we made the decisions in the Steering Committee.

Steering Committee members would participate in a group's discussion of a particular issue until they understood the issue and the options. They would discuss the issues with their national teams and with each other, and finally hold a meeting to resolve them. I usually led the discussion within the Steering Committee and would formulate a resolution that I sensed that Mori and Kadomtsev would support, and then would push for consensus. Grieger frequently resisted but increasingly found himself isolated in the minority. A couple of our Steering Committee meetings ended acrimoniously, but they all ended with a decision that moved the workshop forward.

This procedure caused some disgruntlement and indignant protests from one or two members of the workshop, but it had the desired effect. The topical group members quickly decided that it was better for them to make a decision than to have the Steering Committee make it for them. They were able thereafter, with a couple of legitimate exceptions, to make the necessary decisions to enable the workshop to move forward toward its objective.

Continuing with the example of the Plasma Heating group, it was possible to reject relativistic electron beams because of insufficient evidence that they actually heated plasmas, to reject adiabatic compression because of the requirement for an oversized vacuum vessel that would result in a larger and more expensive device than otherwise necessary, and to reject transit time magnetic pumping because, although it had worked on small experiments, it did not scale favorably to a large device like INTOR.

The case for the electromagnetic wave heating methods was relatively strong, but difficulties were anticipated with the large internal launching structure needed for ion cyclotron resonance heating and with the power sources needed for electron cyclotron resonance heating, and the lower hybrid resonance heating had not yet been demonstrated to heat plasmas. So, in the end, the fact that neutral beam injection heating of a number of tokamak plasmas had been unambiguously demonstrated convinced us that it provided the best assurance for heating the INTOR plasma, despite some technological, size and cost drawbacks, and the need to develop a new type of negative ion source.

Unfortunately, the logic for a choice was not so clear-cut in some of the other groups, in which the advantages and disadvantages for different options tended to be more balanced. I spent a lot of time urging reluctant colleagues to make their best judgment so that we could move on to the next level of detail in our considerations.

* * *

Mori, Kadomtsev, and I all became concerned with the polarization in the Steering Committee and made overtures of various sorts to Grieger. He and I continued our tradition of a dinner together in the Biereklinik and were developing a real friendship.

On another evening, Grieger and I found ourselves walking together down the Kärtnerstrasse to take a taxi to a dinner to which we had both been invited. We were standing at a kiosk buying flowers for our hostess when he said, almost wistfully, that he envied the Americans having the freedom to take initiative on the basis of their own judgment. I thought at first this was a general cultural observation, but quickly realized he meant this not in general but in the context of the INTOR Workshop when he then went on to tell me that he usually took the role of pulling things together that I had assumed in INTOR, rather than the obstructionist role that he was forced to play. (I'm sure that he did not actually say "obstructionist," but that was the gist of his comment.) He went on to tell me about the opposition of Donato Palumbo (head of the EC fusion program) to INTOR and about his own difficulties in dealing with different EC laboratory directors caused by the INTOR Workshop making priority choices among the different technologies or physics operational approaches being developed in these laboratories.

In the future, Grieger and I would make an effort in private discussions to formulate positions that took these difficulties into account, if it was possible to do so without sacrificing the achievement of an objective that, in my view, was essential for the success of the INTOR Workshop. We were not always successful, but a much better working relationship emerged from this enlightening conversation.

*   *   *

The social interactions among the members of the INTOR Workshop contributed to building the trust and friendship among the participants that was such an important achievement of the INTOR Workshop (and that carried forward into the subsequent ITER project). Since the Japanese and Soviets had hosted dinners at Session I, we had arranged for the USA to host a dinner at a restaurant in the Belvedere Park. This had not been straightforward, since there was no budget item for INTOR in the U.S. DoE fusion budget. John Gilleland arranged for Gulf Oil, who owned General Atomics at the time, to pay for this first USA dinner, and he and Paul Rutherford thoroughly enjoyed themselves making the arrangements and selecting the menu, wines, and so on. We had a reception on a beautiful June afternoon in a garden that looked across the Belvedere Park to a Vienna skyline unchanged since the days of the Hapsburgs, and then dinner for some forty or so convivial physicists and engineers in an opulent, gilded dining room. The members of the workshop interspersed themselves among tables with little regard for nationality.

*   *   *

By Session II, the U.S. INTOR delegation had jelled into a smoothly functioning team. In contrast to the hierarchical structure of the Japanese and Soviet teams, in which the members had reported (sometimes several levels removed) to the Steering Committee member in their daily jobs for many years, the U.S. team was a team of peers with much in common. We were all about the same age (40 plus or minus a few years), held doctorates from top universities (Gilleland from Yale; Kulcinski, Wisconsin; Rutherford, Cambridge; and Stacey, MIT) and were modestly accomplished in our field.

Gilleland was a self-effacing but hard-driving practical physicist who had just built the major DIII-D tokamak facility, for which he received the American Nuclear Society's Young Engineer of the Year award. He was very good at making the other participants in any group feel that their contribution was important and valued, so by natural inclination he was effective at moving group discussions forward in the INTOR Workshop.

Kulcinski, a charismatic big man who still looked a bit like the starting guard on the Wisconsin Rose Bowl team that he had been twenty years earlier, had been coleader of the pioneering Wisconsin fusion reactor studies that were the first to explore the characteristics of future fusion reactors. He also had a solid reputation for his work in radiation damage of materials. As it turned out, he was one of the early masters of the art of visual presentation in our field, with slides that gracefully faded from one color into the next, in an era when the rest of us only sometimes had our black and white transparencies typed. Needless to say, he was quite persuasive and a very effective facilitator of group discussions.

Rutherford had made several exceptional developments in plasma theory for which he received the prestigious E. O. Lawrence Award from the U.S. DoE. His ability to organize diverse, inchoate arguments into a set of well-ordered options from which logical conclusions simply emerged was superb. This is a rare quality in any setting, and in the INTOR Workshop it was of enormous value.

As for me, I had some experience in designing reactors for nuclear submarines, had done some theoretical work in both nuclear reactor physics and plasma physics, had organized the multidisciplinary fusion program at Argonne National Laboratory, and had organized the resources of that lab to lead one of the first major U.S. exploratory studies of an EPR. As a result of this rather broad experience, I was able to understand at some level most of what was being discussed at the INTOR Workshop, and I had a rapidly expanding sense of how it all fit together.

We all liked and respected each other. I believe that each of us felt that we were representing our country and our U.S. fusion colleagues. Certainly we each felt a personal, as well as a professional, responsibility for making the INTOR Workshop a success.

During Session II and subsequent sessions, we literally lived, breathed, and dreamed INTOR. In addition to numerous impromptu discussions during the day, we all met together at least every second day to review the progress and the issues that were arising in the various topical group discussions and to develop a U.S. position on them. These discussions frequently continued into the evening over a pitcher of *weisswein* in the dark, catacombic Esterhazykellar or a glass of German champagne in the Reiss Bar, and later over *rostbraten* or *schweineschnitzel* in the restaurant Dubrovnik. Figure 2.2 shows us standing in front of the nearby Stephansdom.

This ongoing discussion among us resulted in the U.S. participants in each of the various groups having a common understanding and position on issues that arose in more than one group, prompting the complaint from at least one workshop member that I was dictating these positions (which may have been possible in some delegations, but not in ours).

*   *   *

One of the charms of Vienna that we discovered early on was the afternoon tradition of the *café conditori,* in which a large fraction of the population of Vienna assembles in a coffee shop cum confectionary between 4 and 5 P.M. to choose among a plethora of confectionary delights to accompany their afternoon cup of black Viennese coffee. My favorite was the Annatorte in the Café Demel, seven layers of chocolate cake and chocolate icing laced with liqueur. The Demel, on the Kohlmarkt between the Graben and the Hofburg, was famous for life-sized confectionary statues of local and visiting dignitaries in the front windows, including U.S. President Jimmy Carter and Soviet Premier Leonid Brezhnev when they met in Vienna during one of the INTOR Workshop sessions in 1979.

*   *   *

My interest in the nearby world behind the iron curtain only whetted by my February trip to Budapest, I spent two weeks getting a visa to Czechoslovakia and then flew to Prague late one Friday afternoon. I had two big meals in cellars on the Old Town Square and shopped the surrounding streets of the old town, without significantly deplenishing the wad of Czech Korunys for which I had exchanged perhaps $100 in shillings in Vienna.

Wandering a block or so off the main squares on Saturday, I came across trucks filled with soldiers, obviously ready to maintain the peace should the need arise. The magnificent old buildings of Prague, which had survived both the Germans and the Russians during the twentieth century, were propped up with poles, and the streets off the main squares smelled of urine. At that time it was possible to walk through the old Jewish cemetery of Prague, with tombstones stacked on top of each other over the centuries, without

Figure 2.2   Zero Phase U.S. INTOR delegation in front of the Stephansdom, Vienna, October 1979. Left to right: W. M. Stacey, J. R. Gilleland, G. L. Kulcinski, P. H. Rutherford.

encountering another soul. (Today, one must wait in line just to view the cemetery through a glass window.)

On Sunday afternoon, I took the train back to Vienna. At the border, the train stopped, and through the window I watched soldiers with Alsatians and submachine guns poke mirrors on long poles under the cars to look for stowaways. Then the soldiers came through the cars, searching the overhead compartments and removing the seat cushions, looking for their fellow countrymen.

A young lout with submachine gun bouncing off a belly barely contained by a partially unbuttoned tunic, a cigarette dangling from his lips, sauntered into our compartment and perfunctorily questioned the occupants, most of whom appeared to be Czech businessmen. When he came to me, recognizably not one, he asked a number of questions in broken English about my business in Czechoslovakia, and then asked if I had any Czech money. Momentarily bereft of my senses and thinking to exchange my remaining wad of Czech currency back into Austrian so that I could have dinner that evening, I pulled it out, at which indiscretion my compartment mates visibly flinched and looked away. The soldier became animated and indicated that I should stay put.

Shortly, he returned together with a very mean looking woman in similar grubby uniform, who asked me in somewhat better English where I got that much Czech money. I told her in Vienna. The two of them then fell into heavy conversation. It dawned on me that they must think I had been exchanging money on the black market (Korunys were "worth" about ten times as much in dollars or schillings at the official Czech exchange rate, which I would have gotten at the airport in Prague when I arrived, as they were on the free market, where I had purchased them in Vienna.) I searched my pockets and bags and luckily came up with the currency exchange receipt from the IAEA, while this mean pair was discussing my immediate future. The woman scowled when I handed her the receipt, they had a further bit of conversation, the beer belly with submachine gun withdrew from the compartment, and the woman wrote out a form attesting that I had a Czech bank account, pocketed my roll of Korunys, gave me another mean look, and walked out. One of my compartment mates said, "You are a lucky one."

When the train rolled forward into Austria, I went to the bar car and had a short celebration of my good fortune at not being incarcerated in a Czech jail for exchanging $100 "on the black market." That was quite enough of the other side for me, and I never went back, until many years later after the Velvet Revolution, when my son went to live in Prague. I never was able to withdraw my $50 from this "Czech bank account," though.

*   *   *

By the end of the four weeks of Session II, the INTOR Workshop had evolved from four mutually suspicious national teams into a single INTOR team. Whereas at the beginning of the session the national teams got together for lunch, by the end of the session the topical groups, with members from each national team, were going to lunch together. We had found a way to make decisions, to compromise, and to combine different national positions into a common INTOR position.

As we sat through the closing plenary session in which the positions of the various topical and special working groups were summarized, it became clear that we were close to agreeing on a reference design concept and a set of reference parameters for an EPR, had assessed the physics and technology database that could reasonably be expected in the next few years upon which to base such a design, had identified and at least partially prioritized supporting R&D to complete this database, and were well on our way to coming to a conclusion that the world's fusion programs were technically ready to jointly undertake such a project as had been proposed by the USSR.

It was also clear that we were well along in building a spirit of international teamwork that would be essential to carrying through such a project. This was exemplified by what remains for me one of those unforgettable moments etched into memory. During the closing plenary session, Vladimir Pistunovich, a giant, bearded Russian physicist, introduced the summary of his group with the words "A divertor, to be or not to be. That is the question." After a moment of silence, the four Americans and two Englishmen in the room cheered, while the other delegates sat mystified. I whispered to Mori, "It's Shakespeare," and he turned to Kadomtsev, who sat on his other side with an equally puzzled look, and said "Shakespeare." By then others had caught on, and Pistunovich towered over us all, beaming. Then everyone clapped, and we carried on with Pistunovich's report, which answered Hamlet's question in the affirmative, at least for the divertor.

By the end of Session II, each of the sixteen topical groups had completed an initial assessment of the status of the physics and technology database in that topical area upon which the design of a major next-step experimental fusion reactor could be based. A preliminary identification of the reference design parameters and performance parameters of such a device had been made, and an initial evaluation of the necessary extensions of the physics and technology database in each topical area had been formulated. This material was documented in INTOR Workshop draft reports that would be reviewed between sessions by experts in the home teams.

The Steering Committee agreed that, in addition to having the home teams review the draft INTOR Workshop reports before the next session, each national team would revise and extend as appropriate their own national reports that had been prepared as input for Session II and publish a national INTOR report summarizing their contributions to the INTOR Workshop. These national reports would be brought to Session III in October, to serve, together with the reviewed draft joint reports that had been prepared at Session II, as a basis for the writing of the INTOR Workshop Zero Phase report at Session III. We also agreed to be prepared to discuss at Session III a recommendation to the IFRC on a next Phase 1 of the INTOR Workshop. The INTOR Workshop members departed Vienna in early July with an optimism that we just might succeed after all in this formidable undertaking, but with a sober recognition that much remained to be done. At this point we had almost become a team.

USA, Summer and Fall 1979

Soon after returning to the USA, I drove up to Oak Ridge to discuss the technical results of Session II with Don Steiner, the head of the recently formed U.S. ETF project, and his chief deputies Tom Shannon and Martin Peng. The general intention of Frank Coffman, the U.S. DoE fusion program director responsible for both INTOR and ETF, was that the material developed for INTOR would be used for the U.S. ETF project as well and, conversely, that the work being performed by the ETF project would also serve as part of the U.S. input to the INTOR Workshop. We determined that in fact the near-term technical work planned for ETF would contribute to the INTOR homework tasks for the next session and that the work being done for INTOR would be useful to the ETF effort that was being planned. The meeting concluded with a briefing for the Oak Ridge fusion program head, Bill Morgan, his deputy, Lee Berry, and two of the leading Oak Ridge plasma physicists, John Sheffield and Jim Callen.

A week later, Steiner and I were in Germantown to brief Kintner, Clarke, Coffman, and other DoE fusion program managers. Coffman reiterated his wish to find a way for ETF and INTOR to work together but to remain separate entities. Steiner and Charles Head, the DoE program manager under Coffman directly responsible for ETF, complained that it was difficult to get people in the fusion community voluntarily involved in ETF because of their involvement in INTOR. Head stated that it would be better for the USA to just undertake ETF and forget INTOR. Kintner indicated that he wanted the U.S. participation in INTOR to continue beyond the end of the year into the next phase (Phase 1).

After the meeting, Frank Coffman took me aside and asked me if I would take over the organization of the ETF activity also. I knew that further expansion of my involvement would require giving up my teaching duties and probably taking a leave of absence from Georgia Tech, which I was reluctant to do. It turned out that he had not yet broached this proposal to Kintner and Clarke, and I felt certain that Clarke would not agree with this plan because of his personal ambitions for the symbolic leading role in an ETF organization. We left it that we would discuss it further after Frank discussed it with Kintner and Clarke. As I expected, this discussion never took place.

* * *

The U.S. INTOR participants (Gilleland, Kulcinski, Rutherford, and myself) met at the University of Wisconsin in Madison to discuss in detail the procedures for reviewing the draft INTOR Workshop reports (which had been distributed to the relevant experts in the U.S. INTOR home team for comment) and the procedures for revising the U.S. topical area assessment reports for publication as a U.S. INTOR report. We also initiated or reorganized special studies of topics that had been identified at the last INTOR Workshop session as high priority: (a) divertors for control of impurity ions in the plasma, (b) determination of the reference major dimensions of the next-step device (which we were now calling INTOR), (c) the poloidal magnetic field coil configuration, (d) the availability of external tritium supplies for INTOR operation, (e) the pros and cons of actually producing electricity (as opposed to creating the fusion conditions from which electricity could be produced using standard technologies), and (f) greater emphasis on materials that could survive in a fusion neutron environment.

In mid-September, the leaders of the U.S. topical group assessments and special studies met with the U.S. INTOR participants and various other fusion physics and technology experts. About fifty of us assembled at Georgia Tech for three days to review and critique both the draft INTOR Workshop reports and the comments on them by U.S. INTOR team members, and also the final draft of the U.S. INTOR report.

Copies of the U.S. INTOR report (consisting of the sixteen topical group reports, seven special topic reports, and a reference concept report, prepared by 122 very knowledgeable people in the U.S. fusion program) coming out of this meeting were assembled into light-blue binders with "US INTOR" emblazoned on the cover for submission to the next session of the INTOR Workshop and for distribution within the U.S. fusion program. The principal authors, in addition to Gilleland, Kulcinski, Rutherford, and myself, were Mohamed Abdou of Georgia Tech; Vic Maroni and Dale Smith of Argonne National Laboratory; Everett Bloom, Jim Callen, John Sheffield, and Jim Watson of Oak Ridge National Laboratory; Alan Dietz, Harold Furth, Dan Jassby, Dale Meade, John Schmidt, Ken Wakefield, and Ken Young of Princeton Plasma Physics Laboratory; Bob Conn of the University of Wisconsin; John Purcell and John Rawls of General Atomics; Neil Young of Ebasco; Jim Crocker of EG&G; and Chuck Flanagan of Westinghouse. The other contributors to the report were from these organizations as well as

MIT, McDonnell-Douglas, Grumman, Sandia Livermore Laboratory, UCLA, Lawrence Berkeley Laboratory, Los Alamos Scientific Laboratory, Brookhaven National Laboratory, Battelle Northwest Laboratory, General Electric, Monsanto's Mound Laboratory, the University of Texas, Bechtel, and Hanford Engineering Development Laboratory.

Paul Rutherford had organized the several assessments of the status of tokamak plasma physics. An assessment of the database for energy and particle confinement in tokamaks, including simulations of implications for INTOR, was led by Jim Callen, head of plasma theory at Oak Ridge. Bob Conn, coleader of the pioneering Wisconsin fusion reactor studies, led the preparation of an assessment of the database on plasma fueling, exhaust, and impurity control. Dan Jassby, the principle fusion exploratory design man at Princeton, led the assessment of plasma heating. John Rawls of General Atomics led the assessment of plasma stability control and the assessment of plasma startup, burn, and shutdown. Rutherford himself led the special studies of the plasma size required for INTOR and of divertors for impurity control, the latter of which included assessments of the poloidal divertor by Dale Meade and Michio Okabayashi of Princeton, of the bundle divertor by John Sheffield and colleagues of Oak Ridge, and of a hybrid divertor by Harold Furth and Christine Ludescher of Princeton and Ted Yang of Westinghouse. John Schmidt of Princeton led the special study on the poloidal field coil configuration.

Jerry Kulcinski had organized the assessments in the materials and nuclear areas. Chuck Flanagan of Westinghouse had led an assessment of the tritium production blanket, the shield, and the plasma chamber first wall. Everett Bloom of Oak Ridge and Dale Smith of Argonne led the team that prepared the assessment of the materials database for INTOR. Vic Maroni of Argonne had led the assessment of tritium recovery and processing systems for INTOR and the special study on tritium breeding in INTOR. Jim Watson of Oak Ridge had led the assessment of the vacuum pumping and containment system. Jim Crocker of EG&G had led the evaluation of safety and environmental requirements for INTOR. Kulcinski himself had led the special studies of materials and tritium-breeding blanket testing requirements for INTOR.

John Gilleland had organized the assessment of the engineering systems for INTOR. John Purcell of General Atomics, who had built the large superconducting bubble chamber magnet at Argonne, had led the assessment of superconducting magnets and the associated cryogenic systems. Ken Wakefield of Princeton had led the assessment of the systems integration and support structure requirements for INTOR. Alan Dietz of Princeton had led the assessment of the power supply and transfer requirements for INTOR. The assessment of the assembly and remote maintenance of INTOR had been led by Neil Young of Ebasco. Mohamed Abdou of Georgia Tech had led the assessment of radiation shielding and personnel access. Ken Young, leader of the plasma diagnostics group at Princeton, had led the assessment of diagnostics, data acquisition, and control requirements for INTOR. Chuck Flanagan of Westinghouse had led the special study of the requirements for electricity production in INTOR. Joe File of Princeton had led the special study of cost, schedule, and manpower requirements for INTOR.

I had organized a concept advisory group to define reference sets of technical objectives and major parameters for INTOR. This group consisted of most of the people in the USA who had been involved in fusion reactor exploratory studies and fusion development scenario studies—Charlie Baker from Argonne, Dan Cohn from MIT, Bob Conn from Wisconsin, Dan Jassby from Princeton, Martin Peng and Lowell Reid from Oak Ridge, John Rawls from General Atomics, and myself.

This U.S. INTOR report was the most comprehensive and broadly based assessment of the status of fusion development made to date in the USA. As such, it was widely used as a reference within the U.S. fusion program for several years. This and other blue notebooks containing reports from later phases of the INTOR Workshop can still be found on the bookshelves of the older generation in the U.S. fusion program.

Similar reports, with comparable lists of authors comprising the leading physicists and engineers in the fusion programs and associated research institutes and industries (see glossary), were being assembled at the same time by the USSR, Japanese, and EC INTOR home teams. The institutions involved in preparation of the USSR report included the I. V. Kurchatov Institute of Atomic Energy, the D. V. Efremov Scientific Research Institute of Electrophysical Apparatus, the Bajkov Institute for Metallurgy, the Kharkov Institute of Physics and Technology, and the All-Union Scientific Research Institute for Inorganic Materials.

Authors of the Japanese INTOR report were drawn from the JAERI, Hitachi, Fuji Electric Co., Kawasaki Heavy Industries, Mitsubishi Heavy Industries, Mitsubishi Atomic Power Industries, Mitsubishi Electric Co., Toshiba Corp., Kyoto University Institute of Atomic Energy, Nagoya University Institute of Plasma Physics, Nippon University, and Nippon Atomic Industry Group.

Contributors to the EC report represented a large number of European research institutions (UK: JET, Culham, Harwell, Warrington; Germany: IPP Garching, KfA Julich, KfK Karlruhe, Technical University of Braunschweig; France: CEA FAR, CEA Fontenay, CEA Saclay, CEN Grenoble; Italy: JRC Ispra, CNEN Bologna, CHEN Frascati, University of Naples; Belgium: CEC Brussels, CEN/SCK Mol, ERM/KMS Brussels; Netherlands: FOM Jutphaas, ECN Petten; Switzerland: CRPP Lausanne, SIN Villigen; Luxembourg: Luxhampton; Sweden: KTH Stockholm, NE Stockholm, NE Studsvik.

Vienna, October 1979

Session III of the INTOR Workshop was held in Prince Eugen's palace on the Annagasse in Vienna October 1–19, 1979. The atmosphere at this session was positive, cooperative, and businesslike from the outset. The workshop had definitely found its stride.

Kadomtsev's report at the opening plenary session reflected a great deal of work in response to the INTOR homework tasks and attention to such things as cost and schedule ($2.9 billion total for design, construction, operation, and decommissioning; ten years for design and construction), manpower requirements, siting and layout of the complex, the testing program, and so forth. This report clearly indicated that the USSR had made up its mind about feasibility and was very serious about moving forward with design and construction.

Similarly, the Japanese plenary session presentation reflected a great deal of work and their usual penchant for detailed design work, as well as input from their industrial sector on detailed planning of schedule and costs ($2.3 billion for construction and operation exclusive of personnel; eleven years for design and construction).

The U.S. plenary presentation also reflected the large amount of work done in response to the INTOR home task assignments, and the U.S. estimate of costs and schedule ($1.33 billion + $200 million contingency, including decommissioning but excluding operating costs; ten years for design and construction).

The EC plenary presentation reflected somewhat less work than the others in response to the homework tasks (everyone in Europe is on vacation in August) and an estimate of cost and schedule ($1.2 billion + $360 million contingency for construction; ten years for design and construction).

Subsequent detailed discussion established that these cost estimates included different items and were actually much closer together than the above numbers would indicate.

Several essential features of the INTOR concept were then discussed in plenary session, so that all workshop participants could be involved in the decisions. There was general agreement that a major radius of the toroidal plasma of 5.0–5.5 m would suffice to provide sufficient plasma confinement while allowing adequate space for the plasma, the tritium-breeding test blanket, the shield, and the magnets, although Folker Engelmann, the EC physics representative, argued for a larger size to provide greater assurance of adequate plasma confinement (but at greater cost).

There was general agreement that some type of divertor would be necessary to magnetically divert ions escaping from the plasma out of the plasma chamber so that their interaction with a material surface would take place in a separate chamber, which would help inhibit the atoms thereby sputtered from the wall from entering the plasma chamber and increasing the radiative cooling of the plasma. There were three divertor concepts, differing in which magnetic field lines were diverted, as shown in figure 2.3.

The more common way at that time to prevent the plasma from interacting with the vacuum chamber wall was to place a "limiter" on the wall (also shown in figure 2.3) that the magnetically confined

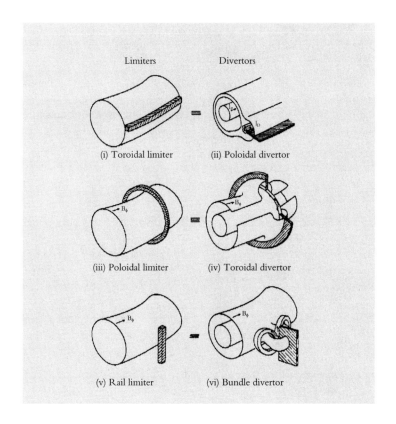

Figure 2.3   Limiter and divertor configurations. (Note that the circular surface represents the plasma surface for the limiter configurations, and the wall if further outside but not shown. For the divertor configurations, both the plasma surface and outside of it the wall are shown.)

ions would strike before they could reach the chamber wall. However, this produced sputtered impurity atoms from the limiter directly in the plasma chamber, whereas with the divertor the sputtered atoms were produced in a separate "divertor chamber" and had less opportunity to get into the plasma.

The INTOR Workshop chose the poloidal (field lines) divertor (figure 2.3, drawing ii), which was being tested in the recently operational ASDEX (EC) and PDX (USA) tokamaks.

The magnetic configuration issue was related to whether at least some of the poloidal field ring coils should be close to the plasma (as the shaping field windings shown in figure 1.1 are) in order to readily control certain anticipated plasma positional instabilities. This would require that the poloidal field coils link with the toroidal field coils (a complicated, Chinese puzzle configuration). Locating all the poloidal coils outside the toroidal field coils led to a simpler configuration but required far more energy for plasma positional stability control. Only the USSR favored the linked coil set configuration, and the INTOR Workshop chose the simpler external ring coil configuration.

Major performance issues included whether or not the INTOR reference concept should (a) actually produce electricity and (b) provide the capability for materials testing in high neutron fluxes for extensive periods of time. It had been agreed that INTOR must provide the capability for the four parties (USA, USSR, EC, Japan) to test tritium–breeding blankets, but (c) whether such tests could be carried out in relatively small loops or intermediate size test modules or whether large blanket sectors were required was an issue. A related issue was (d) whether INTOR should produce the tritium that it used or rely on external sources (i.e., the tritium produced as a by-product in the Canadian heavy-water fission reactors or in the military tritium production reactors of the USSR, USA, UK, and France). These issues were discussed in a plenary session so that all members of the workshop could become aware of their relevance to INTOR.

A subcommittee of the IFRC, the IAEA advisory group to which the INTOR Workshop reported, came to Session III to view the workshop in action and to meet with the Steering Committee. This was not an uplifting visit. The subcommittee members, some of whom surely were apprehensive about how the recommendations in the pending INTOR report might conflict with their national programs and personal agendas, suggested that the INTOR report should be held in confidence by the IAEA and later distributed on a limited basis, and that the cost portion should be published separately. The IFRC chair, Bas Pease, once again rescued the situation (at least for the moment) by steering these concerns and a few off-hand suggestions around to a recommendation that the IFRC receive advanced copies of the report to review and then meet with the Steering Committee in January to discuss the report and continuation of the INTOR Workshop.

Then the workshop got down to the main task of the session, namely, to write the joint INTOR report. Each of sixteen topical groups and the several special groups that had been formed subsequently were responsible for writing their group report, after discussing the review comments on the previous draft reports from the four home teams and after reviewing the relevant sections of the four national INTOR reports. In addition to the four INTOR participants from each party, a number of experts attended Session III to assist with the preparation of the report (EC: G. Casini, J. Darvas, and J. Bohdansky; Japan: H. Momoto and Y. Shimomura; USSR: G. F. Churakov; USA: C. C. Baker and V. A. Maroni).

The Steering Committee asked me to review the various group reports for consistency and to prepare the summary chapter with input from the various groups. I was also asked by the Steering Committee to prepare drafts of the introduction and conclusions chapters. The IAEA provided a technical editor for style and format, but in effect I was to serve as the summary and conclusions writer and technical editor for organization and content on this and all future INTOR reports. I would then edit the summaries somewhat to produce the INTOR Workshop papers published in the IAEA journal *Nuclear Fusion*.

A plenary session on the major conclusions of the sixteen topical groups and the special groups was held in the middle of Session III of the INTOR Workshop. The Steering Committee had previously discussed the need to get the topical groups to reach a conclusion about the adequacy of the database in each topical area to support a design of an INTOR device based on the reference parameters, and to identify specific R&D needed to close any gaps. After a few introductory remarks by Mori, I chaired this plenary session, as was now the customary arrangement.

When the first group leader to report did not make a conclusive statement on this issue of the adequacy of the database, I asked him if, in the judgment of his group, the database that would exist in this topical area within a few years would be sufficient to support the design of a device based on the INTOR reference parameters and, if not, what additional R&D needed to be done to fix the problem. (This is a difficult type of question for a physicist to answer, because while there may be a large possibility that the answer is yes, there is always a small possibility that the answer is no, and physicists are trained to pay attention to small possibilities.)

I persisted with the question in the face of various qualifications and rising irritation until I got something I said that I would interpret as a "yes" if the speaker agreed, which he did. A milder variation of this performance was repeated with the next group leader to speak, who of course was expecting the question, and subsequent speakers addressed the issue without being asked. I can't say that I made a lot of friends on this day, but when the final reports were turned in, this central issue of the Zero Phase INTOR Workshop had been explicitly addressed.

* * *

Years later Tazima, the Japanese physicist during the Zero Phase, escorted me from a meeting in Tokyo to the Plasma Physics Institute in Nagoya to give a seminar. We were leaning back on the bullet train reminiscing about INTOR and enjoying the view of Mt. Fuji over the rice paddies when Tazima confided that at first the Japanese INTOR participants had been upset at the way that I led the workshop by "making plans and telling people what to do," rather than just approving plans made by the people doing the work, as was the Japanese style of leadership to which they were accustomed. He said that Mori had reassured them that what I was doing was the Western way of leadership and that it was necessary in an international workshop in order to accomplish the work that must be done on time. Tazima acknowledged that in the end the members of the Japanese team agreed.

* * *

The final report of the INTOR Workshop Zero Phase was prepared and delivered to the IAEA for technical editing and readying for printing, subject to a review of the final version by the Steering Committee in December. The report contained the following conclusion and recommendation, prepared by the Steering Committee and approved by the INTOR Workshop in plenary session.

Conclusion:
A substantial physics and technology database for INTOR exists today, and this database will be expanded over the next

few years by currently planned programmes. However, certain crucial information will not be developed by the currently planned programmes. Much of this missing information could be developed on the INTOR time scale by the expansion and/or acceleration of existing R and D programmes and by the establishment of new R and D programmes. On this basis, it is concluded that it is scientifically and technologically feasible to undertake the construction of INTOR initially to operate about 1990, provided that the supporting R&D effort is expanded immediately to provide an adequate database within the next few years in a number of important areas. Furthermore, it is concluded that the construction of an INTOR-like device to operate in the early 1990s is the appropriate next major step in the development of fusion power.

*   *   *

Recommendation:

The INTOR Workshop participants recommend that the INTOR project be taken into the next phase—definition. The definition phase should have as its objective the development of a single INTOR preconceptual design and such other material as is necessary to allow a decision to be taken to proceed into the design phase.

Vienna, December 1979

The Steering Committee returned to Vienna for a few days in mid-December to review the proofs of the Zero Phase INTOR report, which were approved for a limited advanced distribution to the IFRC members, and to plan our presentation to the IFRC in January. A photo taken of the Steering Committee meeting is shown in figure 2.4. A decision was also taken to publish the introductory and summary chapters as a paper in the IAEA journal *Nuclear Fusion,* pending IFRC approval of the report in January. We made tentative plans for the next, definition, phase of the INTOR Workshop, again pending IFRC approval, and went home for Christmas.

Figure 2.4 INTOR Steering Committee, Annagasse, Vienna, December 1979. Left to right: Boris Kadomtsev, Sigeru Mori, Bill Stacey, Gunter Grieger.

Vienna, January 1980

The IFRC meeting held January 16–18, 1980, at the IAEA headquarters in Vienna was opened by Director General Sigvard Eckland, who welcomed all participants, praised the work of the Zero Phase of the INTOR Workshop, and indicated that the IAEA was prepared to continue the workshop if the member states so desired and provided the necessary support. In his opening remarks, IFRC Chairman Bas Pease also praised the INTOR report and expressed his opinion that the IFRC will support the recommendation of the INTOR Steering Committee.

The Steering Committee then made its formal report to the IFRC. Mori began by summarizing the framework, major results, and conclusions of the INTOR Workshop Zero Phase. Then I described the INTOR concept, its envisioned role in the world fusion development program, and the technical objectives and probable physical characteristics. Kadomtsev and Grieger followed with summaries of the physics

Figure 2.5 INTOR Steering Committee reporting to the IFRC, IAEA, Vienna, January 1980. Left to right: S. Mori (partly hidden), W. M. Stacey, and B. B. Kadomtsev (not shown: G. Grieger).

and engineering assessments, respectively. A photo taken of the Steering Committee reporting to the IFRC is shown in figure 2.5.

At this point, Pease asked for a straw vote on the Steering Committee recommendation to extend the workshop into Phase 1, the "definition phase." To my absolute amazement, Donato Palumbo (EC), who was sitting to Pease's left and was the first one called upon, responded yes that he agreed, although "there were some questions to be looked into in the definition phase." Ed Kintner (USA) responded that he thought "the objectives and physics were fuzzy." Yevgeny Velikhov (USSR) responded "yes." Husimi (Japan) responded that he was "concerned about the effect of uncertainties." Von Gierke (Germany) responded "yes." Trocheris (France) responded that "it would be useful to continue into the definition phase, even if we stopped at the end." Braams (Netherlands) responded that "he supported Palumbo and that INTOR should carry the tritium-breeding blanket as a backup." Lehnert (Sweden) responded "yes, provided that more information was forthcoming from the existing experiments." Watson-Monroe (Australia) responded "yes." The future ITER had made it over the second major hurdle (the first one having been the decision by the IFRC to form the INTOR Workshop).

The INTOR concept and technical objectives were then discussed by the IFRC. The majority opinion was that INTOR should primarily serve as an engineering test facility for components and perhaps materials testing, although Husimi (Japan) urged that INTOR be the maximum feasible step toward a commercial power reactor, not just a test facility. Velikhov echoed the Steering Committee position that INTOR should demonstrate the physics and engineering components needed for a reactor and serve as a test facility for tritium-breeding blankets, but that electricity production was not necessary because that technology was well known. The IFRC agreed to take responsibility for assuring an external supply of tritium.

The IFRC endorsed the above conclusions of the INTOR Workshop Zero Phase, agreed to ask the IAEA to host Phase 1 of the INTOR Workshop with the objectives and mode of working as recommended by the INTOR Steering Committee, and invited the members of the Steering Committee to continue to serve in their present capacity, including the chairman and vice chairman. The publication and distribution to the governments of the Zero Phase INTOR report were also endorsed. On this positive note, the Zero Phase of the INTOR Workshop concluded.

The 650-page report of the Zero Phase of the INTOR Workshop published by the IAEA in early 1980 as STI/PUB/556 was arguably the most comprehensive assessment of the status of fusion development, perhaps of the status of any scientific program, ever undertaken. In excess of 500 physicists and engineers in Europe, Japan, the USSR, and the USA contributed to the assessment. The principal compilers of the assessment and authors of the report were the INTOR participants—G. Grieger, F. Engelmann, R. Hancox, D. Leger, and P. Reynolds of the EC; S. Mori, T. Hiraoka, K. Sako, and T. Tazima of Japan; B. B. Kadomtsev, G. E. Churakov, B. N. Kolbasov, V. I. Pistunovich, and G. F. Shatalov of the USSR; and W. M. Stacey, J. R. Gilleland, G. L. Kulcinski, and P. H. Rutherford of the USA. (A complete list of authors and content is given in appendix C.) Members of the IAEA Scientific Secretariat, which supported the INTOR Workshop during the preparation of the report, were J. A. Phillips (USA), R. A. Ellis (USA), and V. S. Vlasenkov (USSR). A summary of the report was published in the IAEA journal *Nuclear Fusion* (vol. 20, p. 34, 1980).

Wide distribution of the INTOR report in spring 1980 evoked a generally positive reception among people working in the field and among the authorities responsible for fusion in the various countries, because it documented the status of fusion physics and technology and the readiness of fusion to take a major step in its development. The report also served an important function in focusing the efforts of the world's fusion laboratories, whose leaders wanted to be working on the most important problems in the development of fusion, since an international consensus on these problems was identified in the INTOR report. This was perhaps the high-water mark of the INTOR Workshop.

While the authorities with responsibility for fusion in all the countries involved in INTOR were now in agreement about the desirability of moving forward with a next major step of the sort identified in the INTOR report, opinions still differed regarding how to go about this. The Soviet fusion program, which apparently was already feeling some of the financial constraints that contributed to the downfall of the USSR a few years later (their large tokamak, T-15, was delayed and never operated at full capacity), definitely wanted to be a part of such a major next-step device constructed and operated internationally. The Japanese definitely wanted to be involved with the USA, for whose scientific and engineering capability they still had a great respect at that time, in any such new and complicated venture. The EC (with its evolving political structure and diverse national fusion authorities) had not really thought through the building of such a major next-step device before the INTOR Workshop forced the issue upon them, and consequently did not yet have a consensus position.

The U.S. fusion authorities had given a lot of thought to the topic, but as described above, Ed Kintner, head of the U.S. DoE fusion program office, wanted to build his own version of INTOR as a U.S. project, now called the Engineering Test Facility (ETF). My impression is that as far as he was concerned, the INTOR

Workshop had served its purpose by providing the technical justification for building a U.S. ETF. He had no intention for INTOR to go forward as a design and construction project, even the remotest prospect of which might undermine higher echelon DoE and congressional support for his ETF proposal, although he was circumspect in his statements on these matters to his fellow IFRC members.

The first technical step in the long process leading eventually to ITER had been taken: a detailed technical assessment had concluded that it was possible to undertake the design and construction of an experimental fusion power reactor based on the tokamak. Perhaps equally important, the scientists and engineers from Europe, Japan, the USSR, and the USA who would carry out this design had learned to work together with confidence in each other. Now, the next job was to develop a first conceptual design of an EPR.

3

Phase 1 of the INTOR Workshop (1980–81)

In Phase 1, the "definition phase," the principal objective defined by the International Fusion Research Council (IFRC) for the INTOR Workshop was the "production of a conceptual design supported by a report."

Whereas in the Zero Phase an approximate identification of the probable features and parameters of the INTOR experimental reactor sufficed to guide an assessment of the physics and technology "database" (the status and the achieved parameters of the underlying physics and technology) for the design of such a reactor, it was now time to identify specific materials, dimensions, and technologies for all components and to carry out self-consistent calculations to ensure that given choices of materials and dimensions were mutually self-consistent and would indeed lead to a design of a reactor that was likely to achieve operating parameters within an acceptable range.

"Conceptual design," while sufficient to establish feasibility and identify key problem areas and R&D requirements in detail, stops far short of the level of detail required for the development of blueprints showing the location of holes for attachment bolts, and so on, that is needed for component manufacture. The "engineering design phase" that would be proposed for Phase 2B of the INTOR Workshop was intended to provide a transition to a much more extensive "detailed design phase," which in turn would lead to the start of procurement and construction, but we were not looking that far ahead at this point.

USA, Winter 1980

Since conceptual design would require performing new calculations and analyses in response to specific guidelines, as opposed to assessments of the state of the art for various areas of physics and technology that were performed in the Zero Phase, it was necessary to involve people who were experienced with the necessary codes and, equally important, were supported to devote substantial amounts of time to making such conceptual design calculations in the INTOR work for Phase 1.

Again, this was a relatively straightforward matter for the USSR and Japan, whose INTOR teams were headed by the same people who had authority over their national fusion programs. The Japanese also made extensive use of their industries for the more engineering-oriented calculations. However, it was far from a straightforward matter for the European Community (EC) and the USA.

The U.S. Department of Energy (DoE) fusion program had recently organized an Engineering Test Facility (ETF) design team under Don Steiner at Oak Ridge, and the EC was now in the process of organizing a similar Next European Torus (NET) team under Romano Toschi (Italy) at the Max-Planck-Institut für Plasmaphysik (IPP) near Munich. While both of these teams were assigned responsibilities for designing national experimental power reactors (EPRs), they were also assigned responsibility for supporting the INTOR conceptual design activity. The distribution of effort between the national (ETF or NET) and the international (INTOR) activities was left to be worked out in both instances. Despite the best of intentions and strong personal regard (Steiner and I had shared an office in graduate school at MIT), the potential for conflict was large in such situations.

Reorganization of the U.S. INTOR team for the conceptual design phase was a major task for me during the first half of 1980, requiring much time on the phone and in meetings working out who was to do what for INTOR and how the INTOR tasks could be used in the U.S. national ETF design, and vice versa. Ironically, the problem was enormously complicated by the fact that there was now a budget for the U.S. ETF/INTOR project, and every lab wanted its share of it.

There had been no INTOR budget for the Zero Phase, and all of the Zero Phase INTOR work had just been done voluntarily by people in the U.S. fusion community who were convinced that it was important. Equally important, their DoE fusion program managers had been willing for them to spend time on it.

Now that ETF was an item in the U.S. fusion budget and the responsibility of a given DoE fusion program manager, other DoE fusion program managers with responsibilities for other items in the DoE fusion budget began discouraging "voluntary" work on ETF (and, by association, INTOR).

Chuck Flanagan and Tom Shannon, who led the engineering work at the ETF Design Center (ETFDC), became involved in the INTOR engineering work at this time. They took over the U.S. INTOR engineering responsibility from John Gilleland, who had been asked by John Clarke (deputy director of the DoE fusion program office) in early 1980 to take on the responsibility for coordinating the industrial and fusion laboratory contributions to the U.S. ETF effort.

Two of the major physics-related design tasks—the poloidal field coil system and the poloidal divertor—were being worked on by John Schmidt's group at Princeton, and John was the logical successor to Paul Rutherford as the U.S. physics representative for INTOR in early 1980. As the leading U.S. plasma physics laboratory, Princeton had somewhat more latitude in the distribution of the effort of people working there, and others at Princeton continued to voluntarily contribute their efforts to INTOR as they saw fit.

Argonne National Laboratory and the University of Wisconsin, where most of the expertise in the nuclear area resided, had been playing leading roles in the INTOR work, and I was trying to find a way to keep them involved. However, this was an area in which Don Steiner was interested and wanted the new ETFDC at Oak Ridge to be involved. I was told by DoE that the ETFDC would be the resource available for the INTOR nuclear work.

*   *   *

A meeting was held at Georgia Tech in early March to review the major design issues that we wanted to discuss at the first session of the INTOR Workshop Phase 1. The Participants were John Schmidt, Paul Rutherford, and Paul Reardon of Princeton, John Gilleland of

General Atomics, Chuck Flanagan, John Sheffield, Tom Shannon, and Martin Peng of Oak Ridge, Don Kummer of McDonnell Douglas, Jerry Kulcinski of Wisconsin, Charles Head of DoE, and myself.

First among these major issues was the divertor, a relatively new concept for preventing impurity atoms (which are "sputtered" from the wall surrounding the plasma chamber) from reaching the plasma. This concept was being investigated in the recently operational PDX (USA) and ASDEX (EC) tokamak divertor experiments at Princeton and IPP Garching, respectively. The divertor was receiving considerable attention in the INTOR Workshop.

There was strong interest in the USA in two different types, the poloidal divertor (being investigated on PDX and ASDEX), in which the weaker poloidal magnetic field lines are diverted, and the bundle divertor, in which a small bundle of the stronger toroidal field lines are diverted (see figure 2.2). (Toroidal and poloidal refer to the long and the short, respectively, ways around the toroidal—donut-shaped—plasma chamber.) The poloidal magnetic field was weaker, hence easier to divert, than the toroidal magnetic field, but the geometry of the toroidal divertor on the outboard side of the plasma (see figure 2.2) was easier to accommodate in a design than the geometry of the poloidal divertor underneath the plasma.

We identified several other overarching design issues that needed to be discussed early in the Phase 1 sessions in Vienna. The design of the poloidal and toroidal magnetic field coils, the location of the vacuum chamber boundary, the size of the blanket test modules, the configuration of the structural support system, the procedure for remote component replacement, and so forth, needed to be decided early in the design process in order to fix the overall configuration.

Obtaining an estimate of the likely availability (i.e., the fraction of time in the year that a plasma could be fully operational) of INTOR was important to reconciling the design with the objectives that had been specified by the IFRC. The design concept for dealing with the flux of energetic ions and electrons escaping from the plasma that would be incident on the first wall and target plate in the divertor was a major concern. Also identified as priority issues were the needs to develop standards for the design process and to establish a mechanism to ensure that safety considerations influenced the design process for INTOR from the outset.

These various issues identified at the meeting were then assigned to the appropriate U.S. INTOR Participants for follow-up.

Vienna, March 1980

The January IFRC/Steering Committee meeting in Vienna had been designated Session I of the INTOR Workshop Phase I. Session II met March 24–28, 1980, in the new IAEA International Conference Center in Vienna. We were no longer in our cozy palace in central Vienna, but now several miles away outside the central city in the new soaring glass and steel IAEA towers located out beyond the famous Viennese amusement park—the Prater—on the banks of the Danube. There were a few new faces among the INTOR Participants: EC—G.-P. Casini, F. Farfaletti-Casali, and P. Schiller (Italy) and A. Knobloch (Germany); Japan—K. Tomabechi, N. Fujisawa, and M. Sugihara (JAERI); USA—C. Flanagan and T. Shannon (ETF) and J. Schmidt (Princeton); USSR—D. V. Serebrennikov (Efremov Institute). There were also a few departures of Zero Phase Participants: T. Tazima (Japan) and John Gilleland and Paul Rutherford (USA). The four teams had prepared lists of items that they felt should receive early attention in the conceptual design definition, Phase I.

The purpose of this session was to organize in detail the conceptual design activity of Phase I. We were confronted with the task of designing a device that we estimated would take roughly ten years to design, procure, and construct. Yet, within the next few years the upcoming generation of large tokamaks (TFTR, JET, JT60, T-15) would operate and provide new physics information. Furthermore, the required engineering R&D that had been identified in the Zero Phase assessment would require several years to complete, so new information on engineering technology would also be available in this time period.

In other words, we were confronted with the fact that what we knew about the physics and technology would surely change over the next five years, but we could not simply wait until the new database became available. Our plan was thus to base the design initially upon our Zero Phase projection of the physics and engineering database that was anticipated to exist in about five years (i.e., to make an educated guess what we would learn in the next five years and add it to what we knew then), to continually review the ongoing physics and engineering R&D over this time, and to make design revisions as appropriate and as consistent with a schedule of initial operation in the early 1990s.

The Steering Committee fine-tuned the organizational structure that had been previously discussed for the INTOR Workshop Phase 1 and selected the group leaders so as to achieve a roughly equal distribution among the four Parties while still achieving strong technical competence in these leadership positions within the workshop. The plan was for a Design Coordination Board reporting to the Steering Committee to be responsible for the detailed coordination of the design process. This board eventually consisted of Vladimir Pistunovich (USSR), Ken Tomabechi (Japan), Gunter Grieger (EC), and myself. Reporting to the Design Coordination Board were a Physics Group led by Folker Englemann of the Netherlands, an Engineering Group led by Tom Shannon of the USA, a Nuclear Group led by T. Hiraoka of Japan, a Testing Group led by Gely Shatalov of the USSR, and a Materials Group led by Jerry Kulcinski of the USA.

The Physics Group was assigned responsibility for six physics design analysis tasks: (1) plasma stability and transport, (2) plasma burn cycle analysis, (3) divertor and impurity control, (4) magnetic configuration requirements, (5) heating and fueling requirements, and (6) plasma performance analysis.

The Engineering Group was assigned responsibility for eight engineering design tasks: (1) magnetic systems, (2) remote maintenance systems, (3) vacuum systems, (4) heating and fueling systems, (5) mechanical configuration, (6) structural support systems, (7) heat removal systems, and (8) systems integration.

The Nuclear Group was assigned responsibility for the design of the (1) shielding and (2) tritium systems. The closely related Testing Group was assigned responsibility for (1) design of the blanket test modules and (2) definition of the blanket testing program. Testing of various tritium-production blanket concepts in modules to be provided by the four parties was one of the principal objectives of INTOR.

The Materials Group was assigned responsibility for selecting and qualifying all materials. Other groups for planning, safety, and the auxiliary "balance of plant" systems exterior to the tokamak were identified but not staffed initially.

The homework tasks defined for the next session in June were to carry out specific analyses and calculations that would allow decisions to be made both on the initial values of the major parameters of the INTOR device and on the major technical and programmatic issues that would affect those parameters (at least those that could be resolved at the technical level). The homework tasks were also intended to identify the programmatic issues that must be resolved at the government fusion program management level of the parties (e.g., whether INTOR must be capable of providing its own tritium or could rely on an external supply). In summary, the homework tasks were intended to provide the basis for the detailed definition of the conceptual design tasks for the following six months.

USA, Spring 1980

An intensive effort on the INTOR homework tasks was carried out at Princeton, Argonne, Wisconsin, the ETFDC at Oak Ridge, and other institutions in the U.S. fusion program during April and May 1980, in preparation for the major INTOR Session III scheduled for June. Similar efforts were ongoing in Europe, Japan, and the USSR. The coordination requirements were now greater than in the Zero Phase because analyses and calculations needed to be performed on a consistent basis so that they would come together in the end to form a consistent set of results that could define the starting point for an international engineering design process.

We realized that options would have to be sorted out early in the INTOR design process, so they needed to be well enough defined by the next Vienna session in order to get a fair consideration by the workshop. We were also aware that understanding the effects that a choice of a design option for a given component would have on other components and systems was important in order to avoid having the overall design go in a direction that would later lead to a dead end because of incompatibility among systems. This is a hard enough task when all the people involved are located in the same room, and quite another matter when they are spread across the country and around the world.

Coordination of the U.S. effort was accomplished by telephone, by several visits by the INTOR Participants to the different locations at which the work was being carried out, and by a coordination meeting among the U.S. INTOR Participants (Flanagan, Kulcinski, Schmidt, Shannon, and myself) in early May (this was before the day of instant electronic transmission of documents and drawings that enabled the ITER project to function from three sites a decade later).

A three-day INTOR review meeting was held at Georgia Tech in early June in which the various people performing the calculations and analyses summarized their results for the INTOR Participants and selected experts. Don Steiner (ETF), John Sheffield (Oak Ridge), John Gilleland (General Atomics), Paul Rutherford and Paul Reardon (Princeton), and Don Kummer (McDonnell Douglas) were present as expert reviewers for all three days.

The first day was devoted to a review of the analyses of (1) concepts for the first material wall facing the hot plasma; (2) heat sinks for the thermal energy coming from the plasma ("limiters" and "divertor target plates"), performed at the ETFDC in Oak Ridge; and (3) designs for a lithium-containing blanket that would surround the plasma and capture neutrons to "breed" tritium to replenish tritium burned in the fusion reaction, performed at Argonne. Charlie Baker and D. K. Sze of Argonne attended as additional expert reviewers.

Engineering was the topic of the second day. The analyses at the ETFDC on the mechanical configuration, toroidal field coils, location of the vacuum vessel, estimated reliability of components and resulting availability (percent available operation time) of INTOR, test modules, remote maintenance, design guidelines, and safety were reviewed. Wayne Reierson (ETFDC) informed us that, based on the limited data available on the reliability of components similar to those that would be used in INTOR, the best estimate was that the device would be available for operation 36% of the time (being

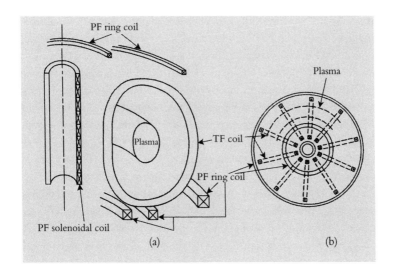

Figure 3.1   Magnet system configuration of the type adopted for
INTOR: (a) vertical cross section showing only a few PF coils, and (b) top
view showing only 2 PF coils (b). PF, poloidal field; TF, toroidal field.

down for maintenance or otherwise unavailable for the other 64%).
Igor Sviatoslovsky (Wisconsin), John Purcell (General Atomics),
Bruce Montgomery (MIT), and Ken Wakefield (Princeton) attended
as additional expert reviewers.

The third day was devoted to a review of physics-related issues.
Martin Peng (ETFDC) reported the analyses of different magnetic
coil configurations for producing the required poloidal field configuration that had been performed at ETFDC and Princeton. The type
of coil configuration eventually chosen for INTOR is shown schematically in figure 3.1.

John Schmidt (Princeton) reported on analyses of (1) the performance of the poloidal divertor for the control of impurity atoms
sputtered from the wall and (2) different "scenarios" for plasma operation (sequence of increasing the plasma current, heating and fueling
of the plasma, increasing the magnetic fields produced by the various
poloidal field coils, etc.) that had been performed at Princeton and
Oak Ridge. Bruce Montgomery (MIT) described calculations of an
alternative "bundle" divertor for impurity control that had been

made at MIT, Wisconsin, and Oak Ridge. Bob Conn (Wisconsin) described analyses of another alternative for impurity control, the "pumped limiter," that he and Mike Ulrickson (Princeton) had made. Bob Miller (General Atomics) reported his calculations of the probable characteristics and impact of large-scale disruptive plasma instabilities in INTOR, and John Rawls (General Atomics) described his analyses of controlling the pulsed operation of a plasma with internal fusion heating—"burn control"—in INTOR. Ron Parker and Dan Cohn (MIT), and Jim Callen, Dieter Sigmar, and Garrett Guest (Oak Ridge) attended as additional expert reviewers.

The U.S. INTOR Participants met following the review and agreed on revisions that needed to be made to the reports just reviewed before their submission to the next session of the INTOR Workshop. We also discussed and decided on the positions that would be taken by the U.S. team on several of the major technical issues that were anticipated to arise at the next session. The major decisions were to recommend the poloidal divertor for impurity control and to recommend the location of all poloidal field coils outside the toroidal field coils, as illustrated schematically in figure 3.1.

Vienna, June 1980

Session III of the INTOR Workshop Phase 1 was held in Vienna in the new IAEA International Conference Center on the Danube, June 16–27, 1980. The Participants were the same as in Session II. The U.S. Participants to Session III are shown in front of the IAEA International Conference Center in figure 3.2.

In his opening plenary session presentation, Gunter Grieger (EC) indicated that now, following the publication of the Zero Phase INTOR report, there was good political support for INTOR in Europe, which was certainly very different from the situation for the early stages of the Zero Phase. In fact, he informed us, Donato Palumbo, head of EC fusion program, was now very supportive of INTOR because many of the European labs that had resisted his efforts to reorient their research programs were now voluntarily reorienting them in accord with the recommendations of the Zero Phase INTOR report.

Figure 3.2   U.S. INTOR delegation in front of the new IAEA Conference Center, Vienna, June 1980. Left to right: Bill Stacey, Tom Shannon, Chuck Flanagan, John Schmidt, Jerry Kulcinski.

The USSR and Japanese plenary session presentations indicated they had reached the same conclusions as the USA on the adoption of a poloidal divertor for impurity control and on the use of a magnetic configuration with the poloidal field coils entirely external to the toroidal field coils (as indicated in figure 3.1), and the EC presentation did not express a position on these two points, so it appeared that we began this session almost in agreement on the major features of the configuration that were necessary to decide early in Phase I in order to allow the conceptual design to move forward to the next level of detail. The remaining divertor issue with a large impact on the overall configuration was whether to have diversion both at the top and the bottom of the plasma (double null) or only at the bottom (lower single null).

The two other unresolved physics-related issues that most affected overall configuration had to do with "field ripple" and "disruptions." The set of toroidal field coils that surround the donut-shaped plasma chamber abut against each other to form a continuous ring on the inboard side (at the hole in the donut) but are relatively widely separated on the outboard side of the plasma (see figure 3.1). This separation produces a "ripple" in the strength of the magnetic field with position around the donut in the long (toroidal) direction, which is strongest on the outboard plasma surface. Since the ripple variation would affect the motion of the magnetically confined ions and electrons in the plasma, it would have an effect (a detrimental one, as it turns out) on the confinement of these particles in the plasma. Thus, the issue was how large the bore (size) of the toroidal field coils must be in order to reduce the confinement degradation caused by the field ripple to an acceptable level. Increasing the bore of the magnets naturally increased their cost, as well as the cost of other components located outside of them.

A "disruption" is a large-scale plasma instability that transfers both the plasma energy and the plasma current to the surrounding structure, creating both a large heat source and a large electromagnetic force in the structure. The likelihood of such events required that the surrounding structure would have to be significantly overdesigned to withstand disruptions, if this was even possible, unless the disruptions could be prevented.

Other important and related configuration issues that needed to be resolved at this session were (1) the exact location of the vacuum boundary between the plasma and the magnets, (2) the procedure for inserting and removing the test modules (horizontally or vertically), and (3) the overall scheme for assembling and disassembling the entire torus configuration and magnet systems.

We had made an initial, rough guess on the configuration issues at the previous Session II, in order to get started. One of the homework tasks had been to work through the consequences of these guesses to see if they made sense and to come to Session III with confirmation or suggestions for changes.

One of the quantities we had estimated was the major radius from the centerline of the torus to the centerline of the plasma (see figure 3.1)—space that had to accommodate (from inside to outside) the central solenoidal magnet, the inboard leg of the toroidal field coil, the shield, the vacuum vessel, the first wall of the plasma chamber, and the plasma. In the USA, we had quickly discovered that this initial guess of the major radius was too small to include all these components, and we had simply increased the major radius by about 10 cm to a value suitable for accommodating our first estimates of the dimensions of all the included components that had to fit into this space. The USSR and EC teams had also found that the initial guess was too small and had made new estimates based on the sizes they had worked out for the included components. The USA, USSR, and EC had all arrived at quite similar new values for the major radius.

The Japanese, however, had reacted quite differently to the discovery that this initial estimate of the major radius was too small. An insightful moment that I shall always remember arose in an early plenary session on this topic when K. Sako, the Japanese engineering representative, strode to the overhead projector with an armload of transparencies. He proceeded to flash one crowded transparency after another before us.

Slowly, we realized what was going on. He had treated this initial estimate of the major radius not as an estimate but as an absolute requirement and had made a monumental effort to design all the components so that he could squeeze everything in. We were being shown detailed finite-element stress analyses of specially designed thin components. This work must have taken thousands of hours of computer time and hundreds of man-hours of engineering to produce. And now we clearly were not going to use it.

I was chairing the session, and it was my duty to keep things more or less on schedule, but I didn't have the heart to stop this heroic presentation. Sako-san was bent over the old-fashioned transparency projector sweating profusely and slapping one transparency after the next down, with a few quick words of explanation. After half an hour he finished his allotted ten-minute presentation. Several members of the workshop waved their hands, eager attack his work. I thanked Sako for an interesting and thorough examination of the issue and announced a coffee break before anyone could question what he was doing. (I think Sako returned the favor years later during a meeting in Tokyo when he took me to a crowded, smoke-filled room up a flight of stairs near Ueno station where grilled baby octopus, small river shrimp, and a few delicacies that I never identified were served from a huge circular stainless steel bar surrounded by perhaps 100 men eating and drinking and generally enjoying themselves.)

There were also issues that needed to be resolved about the level of performance and testing capabilities for which INTOR should be designed. The broad issues were fairly clear from the IFRC charge: INTOR was to produce a "reactor-relevant plasma" using "reactor-relevant engineering technology" and was to provide a capability for "engineering testing." There was general agreement that a "reactor-relevant plasma" would have to operate stably for long pulses (or possibly steady state) at high pressure. "Reactor-relevant engineering technology" generally meant superconducting magnets, reliable heating systems capable of producing about 100 megawatts of power, structural materials that could withstand at least the lower range of the radiation damage and heat flux levels anticipated in future reactors, and so forth.

The testing capability requirements were not so well defined. It was generally agreed that INTOR would have a capability for "nuclear and materials" testing, but the extent of this capability was an open issue. The crux of the issue was the magnitude of the neutron flux that was required for this testing and the period of operating time for which it would be required in order to achieve meaningful testing results. Scoping calculations indicated that any design that met the requirements for a reactor-relevant plasma and reactor-relevant technology would of necessity have a "natural" value of the neutron flux of about one megawatt per square meter of wall area, or less.

Initial studies indicated that a testing program for tritium-breeding modules to be provided by the parties for testing in INTOR could be carried out in about five years of operational time, which would correspond to a neutron fluence (flux × time) of about five megawatt-years per square meter. Achieving this fluence level over a reasonable test period of twenty years while operating at a neutron flux of one megawatt per square meter would require that INTOR be available to operate 25% of the time. On the other hand, meaningful testing of radiation damage effects in materials samples was generally believed by materials scientists to require a fluence of at least ten megawatt-years per square meter, which would require 50% availability over a twenty-year period.

Achieving 25% availability in a first-of-a-kind facility of this magnitude and complexity was judged to be extremely difficult but perhaps possible, while achieving 50% availability was judged to be impossible without a massive, expensive and time-consuming prior program of component reliability development, and probably not even then in a first-of-a-kind device like INTOR. The other possibility for achieving this high neutron fluence was to significantly increase the neutron flux, which would require increasing the plasma performance beyond the "natural" minimal level of a reactor-relevant plasma. This was considered by most of us to be pushing things too far.

To further complicate matters, the lifetime against radiation damage of austenitic stainless steel, which we were coming to realize was the only realistic option for a structural material, was estimated to be about four megawatt-years per square meter, implying the necessity of replacing the innermost parts (first wall and divertor target plates) of INTOR if it operated beyond this limit. We were coming to realize that only a limited testing program of tritium-breeding blanket modules and preliminary materials irradiation was feasible and that the more demanding irradiation testing of materials under high neutron fluence could not realistically be included.

This difficulty of achieving a sufficiently high neutron fluence for meaningful materials testing, as distinguished from operational testing of tritium-breeding blankets, had been encountered in earlier national EPR studies, but this INTOR analysis of the issue was by far the most extensive that had been performed to date. Not surprisingly, the superficially appealing solution of using INTOR to perform the physics demonstration of reactor-relevant plasma operation and the engineering operational testing, and building another plasma materials test reactor in parallel to do the materials testing, came up for discussion.

The first problem identified with this "dual-device" concept was the identity of the second engineering testing device, which would need to be a reliable, high-performance plasma neutron source. Advocates of the dual-device concept usually suggested that this engineering device could be based on some plasma confinement concept other than the tokamak that could achieve high plasma performance in a small, hence less expensive, device. Unfortunately, we were unable to identify any such other plasma confinement concepts that had achieved plasma performance parameters within a factor of 1,000 of those that had been achieved in a tokamak. Even if such a concept existed, to develop another concept to the same performance level as the tokamak and beyond to provide a reliable plasma environment for materials testing would have been in itself a major, lengthy, and hugely expensive endeavor that seemed to us to be unlikely to be undertaken in parallel with INTOR.

Actual electricity production was another programmatic issue that needed to be resolved. A machine designed to simply produce high-quality heat, which could in principle be used to generate electricity but instead was exhausted as waste, would be simpler and less expensive than one that included the energy conversion technology to actually produce electricity. Since that technology was well established, there was nothing to be learned by actually producing electricity, and the only reason to do so in INTOR was the good public relations of a newspaper headline. The physicists and engineers in the INTOR Workshop were not much impressed with this rationale for producing electricity, but the national fusion program managers on the IFRC were.

Detailed homework tasks were defined to address these and other topics, with an emphasis on calculations and analyses that would allow the workshop members to make decisions at the next session. These decisions would define an overall design concept that could then be investigated at the next level of detail, which is always where the devil lies.

*   *   *

On an interpersonal level, the workshop continued to embody the spirit of teamwork and camaraderie that had evolved in the latter part of the Zero Phase, aided again by a long evening over food and wine in a dark Viennese restaurant, hosted by the Japanese. In Sigeru Mori's welcoming remarks he proposed a toast to the "stamina of international physicists and engineers," to which we all responded warmly.

Gunter Grieger and I resumed our custom of dinner together at the Biereklinik. With the change in the European position on INTOR, and thus the conditions under which he had to work, he was a changed man, volunteering for many of the tasks that came up in the Steering Committee and taking a strong, positive leadership position in the workshop. We enjoyed that dinner and many others over the course of an active friendship that extended beyond the INTOR Workshop. In fact, I later sent my son to visit him at the start of his European trip after graduation from college (a sojourn from which he has not yet returned some eighteen years and two grandchildren later).

*   *   *

I had come to know the central district of Vienna quite well by this time. Since the museums closed at noon on Saturdays in socialist Austria, I was usually there when they opened at 9 A.M. The most famous, and justifiably so, was the Kunsthistorisches Museum on the Ringstrasse, with its full room of Bruegels and with rooms of Titians, Cranachs, Dürers, Rembrandts, Canalettos, and hundreds of others following endlessly one after the next. Kulcinski and I were struck by the Bruegel painting of the construction of the Tower of Babel, depicting quite accurately the various technologies of the Middle Ages being deployed to conflicting ends because of lack of communication among people speaking in different tongues. The implication for our own INTOR activity was not lost on us, and Jerry later amused the workshop by showing a slide of this wonderful painting in one of his plenary session talks.

Dozens of Klimts and Shieles, as well as a few Kokoschkas, were housed in the modern Austrian museum of the Upper Belvedere. A real find was the small museum in the Art Academie on the Schiller-platz, which housed an incredible Hieronymus Bosch painting replete with a host of grotesque fiends feeding assorted sinners into the flames of hell. No one but me seemed ever to visit this magnifi-cent painting on Saturday mornings.

My Saturday afternoons were usually spent exploring back streets or inspecting old bones, reliquaries and Hapsburg remains in the many churches. The Hapsburgs appeased both the Catholics and

Martin Luther's followers by entombing dead emperors' entrails in one church and the disemboweled cadavers in another. A particular favorite of mine was the bone room beneath the Stephansdom, where the upper and lower arms and legs of the poor and the plague victims of an earlier day had been stacked like so many matchsticks, with their skulls placed on top of the endless, chest-high stacks. The similarity in length of the two arm bones and the two leg bones was striking. Mozart was said to have been buried here when he died a pauper and in disfavor.

Brussels, July 1980

Most of the physicists at Session III went to Brussels the following week for the biennial IAEA Conference on Plasma Physics and Controlled Nuclear Fusion Research, which is the major gathering of the international plasma physics community. This year the first day was devoted to a plenary session on the results of the Zero Phase of the INTOR Workshop. Mori described the organization of the workshop, I spoke about the INTOR design concept and objectives, Boris Kadomtsev presented an overview of the plasma physics assessments followed by more detailed talks on these assessments by the physics representatives of the other three teams—Rutherford (USA), Englemann (EC), and Tazima (Japan). Grieger then summarized the technology assessments of magnets, heating, and so forth. The conference attendees showed lively interest in the INTOR findings.

* * *

I had a small gastronomical adventure in Brussels that I still remember distinctly. Brussels was then a city of fine French restaurants, many with enticing displays of oysters and other seafood in the windows, which I enjoyed thoroughly. However, I was determined to try also Flemish food and for three straight nights convinced various friends to trek out to a purported Flemish restaurant that I had been told about, only to end up having another excellent French dinner. Finally, on the third night, we found what by all appearances was a real Flemish restaurant in an outlying district, loud and crowded, with bright lights and pitchers of beer on every table.

The menu, of course, was absolutely indecipherable, so I just pointed to one item on the left page and another on the right page. The first thing to arrive was a large slice of a red blob surrounded by gelatin. Seeing the puzzled look on my face, a jolly gentleman at the next table laughingly conveyed that it was bull's nose, pointing to his own with glee. I was gamely washing this down with beer when the laughter level increased at the surrounding tables and a waiter placed a cow's skull filled with another gelatinous mass in front of me. It turned out that I had ordered calf brains in beer, which I had no choice but to gamely wade through, to the great amusement of all. I understood then why it had been so hard to find a Flemish restaurant, and I have never looked again.

*   *   *

The IFRC met in Brussels, and the INTOR Steering Committee reported on the progress of the definition phase of the workshop and on the ongoing Steering Committee planning for the following phase. The IAEA indicated that they could provide office space for a 200-man design effort beginning in January. The IFRC asked the Steering Committee to outline the support needs for a design phase at the October INTOR Workshop session.

USA, Summer 1980

In discussions with Don Steiner (director of the ETFDC), we concluded that it would be more efficient for Chuck Flanagan to be responsible for all ETFDC engineering work for the ETF, and for Tom Shannon to be responsible for all the ETFDC engineering work for INTOR, rather than both of them having responsibilities within both projects, so Tom's responsibilities for the INTOR work increased substantially. The U.S. INTOR Participants (now Kulcinski, Schmidt, Shannon, and I) met in Atlanta in early September to review the progress of the effort on the INTOR homework tasks for the October session. Work was ongoing at Princeton, Oak Ridge, the ETFDC, Argonne, General Atomics, Wisconsin, Brookhaven National Laboratory, Sandia Livermore, Hanford Engineering Development Laboratory, and EG&G Idaho. It proved impossible to get DoE priorities redirected to continue Wisconsin's participation in INTOR, so INTOR lost the considerable talents of Jerry Kulcinski and his team, and Mohamed Abdou, now at Argonne, succeeded him as the U.S. INTOR Participant with responsibility for the nuclear and materials work.

Meanwhile, other developments within the USA would affect INTOR. The Magnetic Fusion Energy Engineering Act of 1980 had been introduced in Congress. The major elements of this bill were a Center for Magnetic Fusion Engineering and a next-step Fusion Engineering Device (FED) to become operational before the turn of the century. Although not yet defined, the FED was generally considered by people in the fusion community to be a national version of INTOR.

In anticipation of the passage of this act, a new Technical Management Board (TMB) was formed by the DoE, led by John Clarke, the deputy director of the DoE fusion program. Its first meeting was held at DoE headquarters in Germantown, Maryland, on September 17, 1980. Present were John Clarke (DoE), John Gilleland (General Atomics), Lee Berry (Oak Ridge), Harold Furth and Paul Rutherford (Princeton), Bob Conn (now at UCLA), Chuck Flanagan (ETFDC), and myself.

Clarke announced that the name of the U.S. next-step device had been changed from the ETF (Engineering Test Facility) to the FED (Fusion Engineering Device), that John Gilleland was responsible for direction of the FED activities, and that the ETFDC was now the Fusion Engineering Design Center (FEDC).

This was now the fourth change of name in six years for the next major facility in the U.S. fusion program. During the initial studies in the mid-1970s, the device had been named the Experimental Power Reactor (EPR), then changed in the late 1970s to the Next Step (TNS) for a second series of EPR studies, and to ETF about a year before this change to FED.

I started the TMB meeting with a summary of the INTOR work and plans. Clarke indicated that the INTOR work would go on at the FEDC, Princeton, and Argonne and that INTOR would have priority at the FEDC and for other community resources until the FED design activity got under way.

Clarke then asked everyone for their opinion of what the role of the new TMB should be. Furth responded that it should be a more active version of the previous advisory board. Gilleland stated that he, Rutherford, and myself would work out the day-to-day details of the FED and INTOR work. Berry stated that the TMB members should make value judgments and high-level decisions and implement work that is identified as needed in their laboratories. He raised the issue that since Steiner was still managing the FEDC, he did not understand what Gilleland's job was. Conn stated that the TMB should develop the FED concept, get industry involved, help industry achieve plasma physics capability, and develop a few ideas. Clarke stated that the TMB should do the job of the new Center for Magnetic Fusion Engineering until the latter was formed, as well as recommend experimental activities that would be done in the fusion laboratories. I suggested that the TMB could review major technical issues and decisions for FED and INTOR activities. Rutherford explained the role of a Physics Group that he was organizing to formulate concepts and options for FED to present to the TMB and to analyze physics issues raised by the TMB.

Clarke stated that the DoE plan was to organize the Center for Magnetic Fusion Engineering by October 1981 and to issue the request for proposals to build the FED by February 1982. Then there was a discussion of the concept for the FED.

By the time of the second meeting of the TMB a month later, Congress had passed and President Carter had signed the Magnetic Fusion Energy Engineering Act of 1980, which authorized the establishment of a Center for Magnetic Fusion Engineering and the construction of a major next-step FED. The prospects for building a U.S. version of INTOR looked most promising. However, as it turned out, an authorization bill is not the same thing as an appropriations bill, and there was an election coming up.

Vienna, October 1980

Session IV of Phase I of the INTOR Workshop was held in Vienna during October 20–31, 1980. The EC Participants were the same as for Session III. The Japanese INTOR Participants now included Ken Tomabechi, an experienced nuclear engineer who had just taken over a senior division leader position in the fusion program.

I knew Tomabechi from a decade earlier when we both were working on fast breeder nuclear reactors and he had spent a year at Argonne. The U.S. Participants now included Mohamed Abdou replacing Jerry Kulcinski, and the USSR team now also included V. G. Vasil'ev as an expert. The other three teams had hardly changed from the beginning of the Zero Phase, but I was the only remaining member of the original U.S. team. Mori had been promoted to executive director of JAERI and transferred from the fusion laboratory in Naka to Tokyo, but he retained his role in INTOR, which was an indication of the Japanese commitment to the INTOR Workshop.

Because of the detailed nature of the design material brought to this session, the opening plenary session was brief, and the Engineering, Physics, and Nuclear groups began group discussions on the first day. The Steering Committee also met on the first day to begin compiling a list of major issues that must be resolved and decisions that must be made at this session and to begin making a detailed outline for the conceptual design report. The Design Coordination Board (Tomabechi, Pistunovich, Grieger, and myself) that was planned in Session II was implemented to oversee the design process during Session IV.

At a subsequent meeting, the Steering Committee members exchanged information that we had agreed to bring regarding national plans for the development of fusion and probable government responses to a recent letter from the IAEA director general asking of their interest in an expansion of the INTOR activity aimed toward design and construction.

Mori reported that the Japanese official plan called for first completing their large tokamak, JT60, and then building first an ignition (energy breakeven in the plasma) test reactor (ITR) and then an EPR. He indicated that the plan was being revisited in light of the INTOR Workshop, that the ITR would probably be dropped, and that the new completion date for the large Japanese tokamak JT60 was 1984. Discussions within the government of doing EPR internationally (i.e., INTOR) were going slowly. He was uncertain about the Japanese response to the IAEA director general's letter asking about their participation in a continuation of the INTOR activity into a design phase.

Grieger stated that there was no official EC plan for fusion development, but the general opinion was that after their large tokamak, JET, an INTOR-like device would be built. A high-level European review committee would meet in 1981 to consider the entire EC fusion program. He reported an EC interest at the political level in maintaining the momentum of the INTOR Workshop, but a concern about creating a centralized project, and skepticism about construction on an international basis. He felt that a centralized organization for INTOR would not be supported but that continuation of the present workshop mode would. He also reported a strong EC interest both in extending JET and in building an ignition experiment (ZEPHYR).

"Ignition" is the condition where the self-heating of the plasma by fusion is sufficient to compensate for cooling by radiation and transport, so that the plasma temperature is maintained by fusion alpha-particle self-heating alone, without the necessity of "external" heating from neutral beams or electromagnetic waves. Ignition— energy breakeven—had been identified as the ultimate goal in the early days of fusion plasma physics research and at this time was still something of a holy grail for plasma physicists, many of whom wanted to finish the physics research before confronting the engineering challenges of a fusion reactor.

By this time, it was clear to those that had looked into it that the engineering and physics challenges of a fusion reactor were interactive and that optimized reactor plasmas would probably operate slightly below energy breakeven, or ignition. However, the goal of ignition was ingrained in the minds of many plasma physicists, and ignition experiments were less expensive than EPRs, so ignition experiment proposals such as the Japanese ITR and the European ZEPHYR found their way into the plans of most fusion programs at one time or another.

Boris Kolbasov ("little Boris") filling in for Boris Kadomtsev ("big Boris"), who arrived a few days later, stated that the only official plans of the USSR were to complete their large tokamak (T-15) and then build an INTOR-like EPR. T-15 was scheduled to be completed in 1984, but Kolbasov was skeptical (he was right—T-15 first operated in 1988). He indicated that the USSR would respond favorably to the director general's letter and could provide 70–100 man-years annually to an INTOR design activity.

I reported the official U.S. plan to build, after our large tokamak (TFTR), an EPR and then a demonstration electrical power demonstration plant. I also informed them of the recent passage of the Magnetic Fusion Energy Engineering Act of 1980 and the redirection of the INTOR-like national ETF conceptual design activities toward a presently undefined, but probably less technically ambitious, FED. I had to tell them that I was uncertain of the U.S. response to the IAEA letter about continuation of INTOR into a design phase, and that U.S. support for the formation of a centrally located INTOR design team seemed unlikely in the present circumstances.

The Steering Committee met again the following day to review major design issues. We had previously discussed and agreed on the necessity to identify early any unresolved issues that must be decided at this session and to meet with the responsible groups to facilitate these decisions. The issues were now relatively detailed, including (1) the location and maintainability of the divertor channel, (2) reacting (balancing) the overturning forces on the magnets, (3) determining if the reduction in "ignition margin" caused by increasing the toroidal magnetic field by 10% was acceptable, (4) determining whether the tritium-breeding blanket should contain lithium in solid or liquid form, (5) assessing the durability under disruption loading of candidate first-wall materials, (6) determining disruption times and energies, (7) deciding between the use of bellows or breaks to increase the electrical resistance of the first wall, and (8) developing a torus assembly and disassembly procedure that did not require warming up the superconducting magnets, along with a few other items. The group chairmen met with the Steering Committee, who impressed upon them the necessity of making decisions on these items at the earliest possible time but definitely by the end of the session.

The senior management of the IAEA had become very interested in INTOR and now wanted to focus the efforts of its Atomic and Molecular Data, Nuclear Data, and Safety and Environment sections on INTOR needs. The Steering Committee agreed that the INTOR Workshop would review the programs of these IAEA sections and prioritize them relative to INTOR needs, and would establish an INTOR Safety Group to work with the IAEA Safety and Environment Section to assess possible safety issues in the INTOR design.

Plenary sessions were held at the end of the first week and again in the middle of the second week to discuss progress in making design decisions. Four major issues were identified on which decisions were still needed: (1) field strength reduction in the toroidal magnet coils, (2) the testing program, (3) first-wall structural materials, and (4) choice of the lithium-containing material in the tritium-breeding blanket. Decisions were made by the end of the session, the schedule was reviewed vis-à-vis the current status of the design activity, and the date of June 1981 was confirmed for the completion of the conceptual design.

The members of the INTOR Workshop then developed a plan for the future of the INTOR activity leading to the beginning of construction in June 1986. This plan included an INTOR Workshop Phase 2A continuation to produce an updated conceptual design, designated the Reference Design, in June 1982. This would be followed by Phase 2B, with a Central INTOR Design Team supplemented by national INTOR design teams producing an engineering design that would support a decision on construction by December 1984, and the development of a production design for component manufacture by a central team by June 1986. This would enable the start of procurement and construction in June 1986. (A similar plan was finally implemented in the ITER project almost a decade later, but with an extended time scale to accommodate prolonged negotiations, and procurement and construction started in 2009.)

*   *   *

The U.S. INTOR Participants enjoyed Vienna's many good restaurants during our stays (none more than me). In the Zero Phase, the Dubrovnik just off the Graben, which served Austrian and Croatian dishes, was the favorite haunt of the American team. The new team during Phase 1 favored a basement restaurant on the Am Hof, a square near the end of the Graben in central Vienna, where we frequently enjoyed a huge Weiner schnitzel overhanging the plate followed up with a Salzburger Nockerl, an airy and irresistible concoction of meringue and chocolate sauce, also overflowing the plate.

My favorite restaurant over the years was the nearby Ofenloch, which served Austrian specialties including those delicious little *spaetzle* dumplings that come with everything in Germanic countries and a *frittatensuppe* to die for, a chicken broth in which floated thin potato sticks fried to a crisp in pork fat (McDonald's, eat your heart out). A close second was the Hummerbar, a walk-up seafood restaurant near the Staatsoper that was perfect for lobster bisque and gravlax after the opera. The Chez Robert in an outlying district was one of the few places where a variety of good seafood could be obtained (the owner drove a day each way to Paris each week to bring it in). Certainly the finest restaurant that I found was the Mattas in the Fleischmarkt, which served a varying five-course continental menu of distinction.

USA, Late Fall 1980

At this point, after a frustrating year of micro-coordination of the U.S. DoE resources available for FED and INTOR design activities, a workable INTOR design effort seemed at last to be fully in place. The DoE had designated certain resources to support the two design efforts, most significantly the manpower for engineering design at the FEDC at Oak Ridge led by Don Steiner, but also including a team with strong nuclear and materials capabilities at Argonne National Laboratory, supplemented by the McDonnell Douglas engineering team. There had been a general agreement that this manpower would be used for INTOR until the FED design effort got going, and then priorities would have to be set. The FEDC (the new acronym for the design center at Oak Ridge) was also staffing in the nuclear area, and there had been pressure from DoE to use that capability for INTOR rather than redeploying the more experienced Argonne and Wisconsin personnel.

The people working to get the FED design going, of course, thought they needed priority for the FEDC resources from the outset. The result was that it was necessary to negotiate with Steiner and later Gilleland and others for manpower to perform the individual INTOR homework tasks. These negotiations were in great detail. I have records of telephone calls that list 15–20 names, the tasks they are needed for, and the notation that we had agreed to 0.5 or 1.0 or sometimes as much as 2.0 man-months of effort from each of these people to get one set of INTOR homework tasks performed. I have records of other phone calls from the INTOR Participants identifying this or that task that simply was not getting done. Then, once we pulled together this work for the INTOR session in Vienna and came home with new homework tasks, the whole process had to be repeated for the new homework tasks.

Fortunately, this period of micro-coordination seemed now to be almost behind us. I had an agreement with Steiner that INTOR would get about one-third of the FEDC effort on a continuing basis and that all this would be coordinated internally by Shannon for INTOR. I had been unable to get resources redeployed to enable the Wisconsin team under Kulcinski to continue to support INTOR, but had been able to get my old group at Argonne, and their collaborators at McDonnell Douglas, redeployed to INTOR work.

Unlike the engineering and nuclear design efforts, the physics design analysis effort under John Schmidt at Princeton, with support from General Atomics channeled through John Rawls and from Oak Ridge channeled through Martin Peng, had been in good shape all along. The Princeton Plasma Physics Laboratory was the principal U.S. tokamak laboratory, and many of its physicists, including Director Mel Gottlieb and his heir apparent, Harold Furth, supported INTOR, so Schmidt was able to get the support he needed for the INTOR work without much micro-coordination. Likewise, Tihiro Ohkawa, director of the fusion program at General Atomics, supported INTOR, making it possible for John Rawls to bring important additional physics resources into the INTOR design effort.

Soon after returning from Session III, Tom Shannon, Tom Brown, and George Fuller from the FEDC and I went to Argonne to meet with Mohamed Abdou and the new Argonne and McDonnell Douglas nuclear design team. The purpose of this meeting was primarily to get this new team up to speed on the INTOR tasks. We reviewed the nuclear and engineering homework tasks and identified a number of other specialists that could be brought into the nuclear and test program design effort on a consulting basis.

Paris, November 1980

Gunter Grieger and I met with a subcommittee (Bas Pease and Donato Palumbo,EC; Ed Kintner, USA; Yevgeny Velikhov, USSR;

Yamamoto, Japan) of the IFRC in Paris on November 11, 1980. I provided a technical summary of the status of the INTOR design, and the two of us answered technical questions for an hour or so.

The IFRC subcommittee members, representing the governments of the four Parties to INTOR, then indicated the likely nature of their governments' response to the IAEA director general's letter inviting their participation in and support of future INTOR activities beyond the current Phase 1 ending in June 1981. Kintner stated that the USA was ready to discuss the matter. Velikhov stated that the USSR approved continuation of the workshop through Phase 2, and that the money was already allocated. Palumbo stated that the EC had a preliminary position and was awaiting the Phase 1 INTOR report to study before replying. He added that approval of Phase 2B involving the formation of a central team would require approval of the EC Council of Ministers. Yamamoto stated that Japan found the INTOR Workshop Phase 2A acceptable but that the formation of the central team for Phase 2B would require further discussion. He added that a review committee was evaluating the long-term fusion program of Japan.

The IFRC then requested that the Steering Committee discuss the design phase with them in January 1981, providing costs, work to be done, and a description of the functions envisioned for the INTOR central team in Phase 2B, the design phase. The IFRC subgroup members then discussed the formation of a small committee to look into the administrative aspects of organizing such a project as envisioned for Phase 2B of INTOR (a permanent central team supported by home teams of the four Parties). A committee of foreign ministry, legal, and technical people from each of the four Parties, with some input from the INTOR Workshop, was suggested. Their job would be to develop an administrative plan and policy on how to administratively organize Phase 2B. The IFRC subgroup agreed that this plan should be completed by December 1981. It was further agreed that the full IFRC should make such a recommendation at their January meeting to the director general of the IAEA. Bob Ellis, the IAEA (USA) scientific secretary attending the meeting, was tasked to draw up a draft recommendation.

Kintner reported that the Magnetic Fusion Energy Engineering Act of 1980 had passed the U.S. Congress and been signed into law, and that there was $45,000,000 in the 1982 budget for creation of a Center for Magnetic Fusion Engineering.

USA, Winter 1980–81

A meeting of the U.S. INTOR Participants (Tom Shannon, FEDC; John Schmidt, Princeton; Mohamed Abdou, Argonne; and myself) and the FEDC staff was held at the FEDC in Oak Ridge on December 12, 1980, to review progress on the INTOR design homework tasks. Shannon reported that the top priority issues for the engineering design were (1) deciding among the options for the poloidal field magnet coil design configuration, (2) the structural support for the toroidal field magnet system, (3) the electromagnetic design (resistance, conducting paths, etc.) for the torus, (4) the torus assembly procedure, (5) joints and connectors in the magnet systems, and (6) the neutral beam injector systems. Each of these was discussed in detail. Abdou reported on the nuclear system design: (1) the first wall facing the plasma, (2) the tritium-breeding blanket, and (3) the divertor heat flux "target plate." Schmidt concentrated his report on the calculations in progress to confirm the ability to achieve the type of magnetic configuration required to form the poloidal divertor and the identification of the coil locations that would be necessary. We planned a review meeting with external experts in early January.

*   *   *

The third meeting of the TMB, which was charged by DoE with guiding the development of the new U.S. FED and the U.S. INTOR activity, was held in Los Angeles on December 18, 1980. In attendance were John Clarke as chairman (DoE), Don Steiner (FEDC), Bruce Montgomery (MIT), Charlie Baker (Argonne), Chuck Flanagan (FEDC), John Sheffield (Oak Ridge), Paul Rutherford, Wolfgang Stodiek, and Harold Furth (Princeton), Bob Conn (UCLA), John Gilleland (General Atomics), Keith Thomassen (Lawrence Livermore National Laboratory), and myself. Gilleland announced that this meeting was intended to focus on the FED concept and that the next meeting would focus on the mission. Rutherford and Sheffield offered (different) energy confinement scaling laws to be used in the FED design. There was a discussion of the FED concept, which generally was considered by the attendees to be less ambitious than INTOR both in physics performance and in engineering testing capability, but was otherwise not defined.

* * *

The U.S. design analysis work on the INTOR homework tasks for Session V was presented and reviewed by the U.S. INTOR review committee (John Gilleland and Fred Puhn of General Atomics, Jerry Kulcinski of Wisconsin, Don Kummer of McDonnell Douglas, Bruce Montgomery of MIT, Paul Reardon and Paul Rutherford of Princeton, and John Sheffield of Oak Ridge) at Georgia Tech on January 6–8, 1981.

Design analyses of the major engineering systems were presented: (1) torus configuration (Tom Brown, FEDC), (2) toroidal and poloidal magnet systems (R. Derby and Dick Hooper, FEDC), (3) electromagnetics (J. Murray and S. Thompson, FEDC), (4) torus integration (George Fuller, FEDC), and (5) neutral beam injectors (D. Metzler, FEDC).

Physics design analyses on (1) magnetic formation of the divertor (Dennis Strickler, FEDC), (2) divertor physics (Doug Post and Marion Petrovic, Princeton), (3) evolution of the magnetic equilibrium (Martin Peng, FEDC), (4) magnetic field ripple effects (David Mikkelson, Princeton, and Jim Rome, Oak Ridge), (5) plasma startup and shutdown (Dan Cohn, MIT), (6) plasma burn control (John Rawls, General Atomics), and (7) achieving plasma ignition (Wayne Houlberg, Oak Ridge) were presented and discussed.

Analysis of the nuclear design homework tasks on the first wall, divertor target plate, and tritium-breeding blanket were presented by Mohamed Abdou and Dale Smith of Argonne and Dave Morgan of McDonnell Douglas.

Then the review committee met with the U.S. INTOR Participants for a half-day following the presentations for detailed discussion of the work and identification of likely solutions to problems that had been detected in the review.

Vienna, January 1981

Session V of Phase 1 of the INTOR Workshop was held January 19 through February 4, 1981, at the IAEA International Conference

Center in Vienna. In addition to the INTOR Participants, the Japanese brought two experts from industry (S. Itoh, Hitachi, and Y. Sawada, Mitsubishi), the U.S. brought Dale Smith (Argonne) as an expert in the nuclear area, and the USSR brought a draftsman (V. A. Loktev). The opening plenary session presentations by all parties were detailed and extensive, indicating that a great deal of effort had been expended on the design homework tasks by all four teams.

The Steering Committee met on the second day to identify issues and decisions that must be made at this session. In the physics area the issues were (1) the need for 100 volts for plasma breakdown, (2) achievement of a 500-second plasma burn pulse, and (3) the margin of the design above ignition (the condition at which the plasma self-heating just balances the plasma cooling due to radiation and transport losses). The engineering issues were (1) torus segmentation (twelvefold or sixfold symmetry), (2) structural support of the toroidal field coil magnet system, and (3) the configuration of the poloidal field ring magnets (inside or outside the toroidal field magnets). The issues in the nuclear area were (1) use of carbon armor on the stainless steel first wall to protect it from the plasma, (2) the extent of the tritium-breeding blanket (test module or totally surrounding the plasma chamber) and the choice of materials, (3) the materials and design of the divertor target plate (which must withstand high heat fluxes), (4) the size and effect of manufacturing flaws that would go undetected, and (5) the effect of longer plasma burn pulses on materials. From the detailed nature of the issues under discussion, it was obvious that the workshop had been successful in getting beyond the more superficial issues and into the issues that ultimately determine success or failure.

We then discussed the responses of the four governments to the IAEA director general's letter regarding the following phases of INTOR. The gist of the responses was that the governments were prepared to go forward with a continuation of the present workshop mode of INTOR but were not prepared to form a permanent central design team without further discussion.

The Steering Committee agreed that a continuation of the INTOR Workshop while those discussions about forming a permanent design center were being held was important in order to maintain continuity and momentum, and we exchanged views on the type of tasks that might most profitably be undertaken. I was asked to give my views first, which were that the workshop mode of operation would be suitable for a collaborative analysis of critical technical issues identified in the design, for further iteration and refinement of the conceptual design, and for an update of the assessment of the database and required R&D. Gunter Grieger agreed with this and also suggested an assessment of the cost-effectiveness of advancing fusion through the construction and operation of INTOR in contrast to trying to achieve comparable advances through several special purpose facilities that were on a smaller scale. Boris Kadomtsev's suggestions—the definition of facilities, siting requirements, support systems, detail requirements, and so on—reflected the USSR determination to move the INTOR activity forward to a design project as soon as possible. Sigeru Mori indicated his general agreement throughout by nods and short follow-up questions but did not express a separate opinion, undoubtedly glad at last for an opportunity to return for a moment to the Japanese style of management.

The homework results brought to Session V were reviewed and compared in detail in the Physics, Engineering, and Nuclear group meetings. This information proved sufficient to allow most of the remaining decisions affecting the design configuration to be made and detailed calculations to be defined for homework tasks for the next session. Because of the possible impact of the decisions being made on other systems, there were frequent plenary sessions to ensure that all members of the workshop were cognizant of developments. In addition, the national teams met frequently to discuss the evolving design decisions and how they would affect other technical issues. The progress made at this session was sufficient to allow us to confirm the completion date for the conceptual design report as June 1981, with the intent of producing a preliminary version at the next session.

* * *

The IFRC met January 27–28, 1981, in parallel with the INTOR Workshop, so that Steering Committee members could meet with the IFRC. After hearing from the Steering Committee that the INTOR design process was on schedule and that the members of the workshop were generally satisfied with the design that was evolving, the major topic of the IFRC discussion turned to the response of the governments of the four parties to the IAEA director general's letter inquiring about the governments' intentions vis-à-vis continuation of the INTOR activity into the design phase.

1. Ed Kintner (USA) stated that he was disappointed in the Japanese and EC letters and that he felt that before a design phase could be entered the administrative relationships must be worked out. He opined that there was no point in design if it could not lead to construction, so it was essential to study the administrative arrangements for a possible future design and construction project. He added that the USA was prepared to continue any meaningful form of INTOR activity, but that it was better to put resources into national efforts (the EC NET, the U.S. FED) if governments were unwilling to set up administrative arrangements to move forward with an INTOR design phase.

2. Kakahana (Japan) stated that the Japanese position (continue the workshop mode, but further discussion on a central design team was needed) was a "soft" position and that he hoped for a positive change in that position.

3. Yevgeny Velikhov (USSR) indicated that the USSR was ready to discuss building INTOR. He recommended preparing for government discussion of administrative arrangements within the framework of the INTOR Workshop up to the end of Phase 2A, after which he was prepared to speak to his government. He urged the IFRC to work toward the realization of INTOR by July 1982.

4. Donato Palumbo (EC) reiterated that it was necessary to see the Phase I conceptual design report before a decision to continue could be made. He said that EC fusion technology was behind that of the other parties and that the NET project had been formed to focus on technology, but that he was uncomfortable because in the past first priority had been on INTOR. He voiced his personal opinion that the EC would participate in a continuation of INTOR if there was a large chance that it would be built. He concluded by noting that the upcoming high-level European review panel on fusion would have a big effect on any decision, and that setting up an administrative committee prior to completion of that panel was incompatible for the EC.

In addition to Palumbo, who spoke for the combined EC fusion program, the representative of the individual national programs in Europe serving on the IFRC also expressed their opinions.

5. Lehnert (Sweden) made a remark about fourteenth-century physics that I did not quite catch.

6. Trocheris (France) said that uncertainties were larger than indicated, making it impossible to go into the design phase.

7. Bas Pease (UK) indicated that he was prepared to go into the design phase of INTOR.

6. Von Gierke (Germany) stated that his country would not go into the design phase unless there was greater than 50% probability of going on to construction.

The following day, Pease, the IFRC chairman, summarized the above positions and proposed a compromise on the issue of administrative arrangements: (1) each country would nominate two people to serve on a committee; (2) the IAEA would provide a paper on administration; (3) the committee would address only the design phase, with some consideration of the future (here Kintner objected that this avoided the issue); (4) the IAEA director general should write a letter to the governments of the four parties requesting such a committee; and (5) the committee should be an IAEA working party.

Kintner responded that it was necessary to decide that eventual construction is practicable before committing significantly more resources, and added that this was not the same as commitment to construct. He argued that the decision must be addressed by people at the level who can eventually make a construction decision.

This IFRC meeting ended with recommendations to the IAEA (1) to continue the INTOR Workshop through Phase 2A until December 1982, (2) that fusion needs international cooperation, and (3) to establish a committee on administrative arrangements in which the IFRC, among others, would be willing to participate.

* * *

While the INTOR Workshop was taken as extremely serious business by all the Participants and by most fusion scientists and engineers that contributed in the home countries, moments of horseplay and relaxation were important elements of catharsis. Figure 3.3 shows

Figure 3.3   The INTOR Engineering Group perfecting the art of compromise, Vienna, January 1981. Left to right: V. Vasil'ev, USSR; G.-P. Casini, Italy; T. E. Shannon, USA; K. Sako, Japan; T. Brown, USA; F. Farfaletti-Casali, Italy; V. Loktev, USSR; Y. Sawada, Japan.

the Engineering Group reaching a compromise on the mechanical configuration.

USA, February–March 1981

I wasn't home a week when a phone call from Tom Shannon informed me that my earlier delusion that the organization of support arrangements for the U.S. INTOR home team was in good shape had been just that—a delusion. Tom told me that the day before, Don Steiner, director of the FEDC, had taken the position with John Gilleland, director of the FED design for the TMB, that he could not meet both the FED and INTOR commitments. Tom relayed that the agreement finally reached between the two of them was that Tom Shannon and Tom Brown could work on INTOR but that the other ten to fifteen people at the FEDC would work on FED, effectively wiping out much of the INTOR engineering design effort.

My next phone call was to my friend John Gilleland, who defended his action by telling me that Ed Kintner and John Clarke, director and deputy director, respectively, of the DoE fusion program office, had instructed him that they wanted a minimum INTOR effort and minimum utilization of the FEDC resources by INTOR in Phase 2A. I reminded him we had six more months of Phase 1 in which we needed to complete a conceptual design. I also gave him my estimates for resource requirements for Phase 2A, which would require a continuing commitment of Schmidt's people at Princeton and the Argonne/McDonnell Douglas nuclear people, as well as fixing the FEDC problem he had just created for me, which conceivably could be done by expanding the Princeton, Argonne/McDonnell Douglas, and General Atomics contributions to INTOR. Gilleland agreed that some relief was necessary to salvage all of the work that had gone into the INTOR conceptual design already.

A phone call to my old graduate school officemate, Don Steiner, gave me a little immediate relief (allowing some people to complete their immediate INTOR tasks), but not enough.

*　*　*

The TMB met on February 26, 1981. A U.S. election had been held and a new government under President Ronald Reagan had been formed. The new director of the DoE Office of Energy Research, Douglas Pewitt, to whom Kintner now reported, had testified to Congress and also met with Kintner and Clarke. According to Clarke, his testimony had conveyed warm but ambivalent support for fusion, although he indicated that there was no urgency for fusion because of the new administration's policy to expand the use of fission energy. He also testified that the government would study FED but that they were not ready to build it, and anyway the government could not afford a major thrust in fusion in times of fiscal austerity. Pewitt further testified that the new government had no intention of following the Magnetic Fusion Energy Engineering Act of 1980. He had initially told Kintner to halt the TMB but then relented to instructing him to keep it low profile.

*　*　*

A review of the U.S. INTOR conceptual design work was held at Georgia Tech on March 16–17, 1981, followed by a meeting of the U.S. INTOR Participants to coordinate the preparation of the U.S.

INTOR report and to discuss the technical issues raised in the review. Tom Shannon and Tom Brown had made a heroic effort on the engineering design tasks, and the nuclear and physics design analysis presentations were solid.

Vienna, April 1981

Session VI of the INTOR Workshop Phase I was held March 30 through April 10, 1981, in Vienna. In addition to the regular INTOR Participants, T. Kobayashi (Hitachi) and N. Miki (Toshiba) were part of the Japanese delegation. Reports made at the opening plenary session indicated that the design analyses were converging on solutions in all but a few areas, which were identified for intensive attention during the next two weeks. We agreed that the preparation of a complete "skeletal" report (rough draft or detailed outline) was an objective of this session.

The Steering Committee then met and exchanged information on the status of INTOR considerations in their respective national programs.

1. Sigeru Mori indicated that a review of long-range fusion plans in Japan, which identified an INTOR-like device in the mid-1990s costing $3 billion, was a shock to the government. He indicated that he (now deputy director of JAERI) was negotiating with the finance ministry and working to get a "soft" approval that would allow them to enter a design phase of INTOR and was discussing international cooperation on INTOR broadly within the government. He stated that the Japanese had identified two people for an INTOR administrative committee—T. Hiraoka of INTOR and X. Yosizawa, an administrator from the office of international affairs of JAERI.

2. Boris Kadomtsev indicated that the main effort of the USSR fusion program was the completion of construction of the large tokamak T-15 and that the government view on INTOR remained as stated several times by Velikhov—they were ready to continue. He smiled and said "It is a good government position—scientists should develop the base for fusion power." One member of an administrative committee, G. Eliseev of the Kurchatov Institute, had been identified.

3. Gunter Grieger informed us that the EC fusion review committee under Karl Beckerts would complete its report in July and that JET add-ons were becoming a major part of the EC program. He suggested that a brainstorming session rather than a formal committee on administrative arrangements for INTOR would be better.

4. I tried to put a good face on the events in the USA, but my report was a bit bleaker than the others. I had to tell them that the U.S. DoE fusion program was increasing the effort on the U.S. FED, at the expense of INTOR, but that the new government did not appear prepared to build FED.

The Steering Committee met with the Nuclear, Engineering, and Physics groups to facilitate taking final decisions that would allow a conceptual design to be completed. The workshop produced a reasonably complete and consistent conceptual design and skeletal draft of the report by the end of this session. Final confirmatory analyses were identified as homework tasks for the next session.

* * *

There was a further indication during this session of the seriousness with which the USSR took their proposal to move forward with design and construction of INTOR. Yevgeny Velikhov, director of the Kurchatov Institute and head of the USSR fusion program, stopped in for an afterwork INTOR wine and cheese gathering. After making the rounds and shaking hands, he and Kadomtsev asked if they could have a word with me. We stepped to a corner of the room somewhat removed from the others, and they proceeded to make the case for hosting INTOR in the USSR—cheap labor, cheap electricity, cheap construction costs, trained manpower, and so on. Velikhov said that a site on the USSR border could be arranged so that there could be entry into the site for INTOR workers without going through Soviet border security.

This was a surreal experience. It was at the height of the Soviet-Afghan war, and some particular atrocity had been in the headlines for the past few days. I responded that it sounded like an interesting possibility but that this might not be the best time to bring it up with the other governments.

* * *

Another Viennese tradition that I enjoyed was the coffeehouse, where people with time on their hands would sit talking or reading, drinking coffee, and invariably fouling the air with cigarette smoke. Each coffeehouse seemed to attract its own type of clientele. My favorite, because it attracted an interesting crowd and was near my pension Riemergasse, was the Café Hawelka in the Dorotheegasse off the Graben. I frequently stopped in before retiring for the evening for *buchteln,* a house special puff pastry stuffed with apricots and powdered with sugar, traditionally taken with coffee and apricot schnapps.

USA, Spring 1981

The two months between Session VI and Session VII were spent performing confirmatory analyses, reviewing the draft INTOR conceptual design report, and putting the final touches on the U.S. INTOR report, which summarized the U.S. contributions to INTOR Workshop Phase I. Actually, we had published an intermediate report in the summer of 1980, and the present report concentrated on material developed since then. Both of these reports were placed in blue binders embossed with "US INTOR" and took their places on the bookshelves of many members of the U.S. fusion community. Several hundred scientists and engineers in the U.S. fusion program contributed to the preparation of these reports. A technical summary of the U.S. contributions to the INTOR Workshop Phase I was published in the journal *Nuclear Technology/Fusion* (vol. 1, p. 486, 1981).

Vienna, June 1981

The final Session VII of the INTOR Workshop Phase I was held June 22 through July 3, 1981, in Vienna for the purpose of writing the final draft of the INTOR conceptual design report. The Japanese provided a reception to celebrate this event, and we completed the task with only minor difficulties. This was a busy time for me because I was responsible for technical editing the report for content and consistency, and for preparing the introductory and summary & conclusions chapters.

At 850 pages, it was quite a hefty report, which was nicely summarized in the forward written by Hans Blix, the director general of the IAEA at the time:

> The Phase-One report of the International Tokamak Reactor (INTOR) Workshop is the culmination of an international effort that required approximately 120 man-years of work. The report, which documents the conceptual design of the next major fusion experiment beyond those being constructed at present, establishes the basis for the next phase of INTOR, the detailed design of the machine. I take pleasure in thanking the Workshop Participants, as well as the members of the IFRC, whose efforts have made this important example of international co-operation in the peaceful uses of atomic energy possible.

Alas, the detailed design was not to come as soon as the director general and all of us hoped. It would require political intervention at the highest level and more than a decade.

Altogether, about 500 scientists and engineers from the four parties contributed to the preparation of the Phase 1 INTOR conceptual design report, which was published by the IAEA as STI/PUB/619 in 1982. This was possibly the first truly international conceptual design of a major scientific facility of this scope. The principal authors were the INTOR Participants, with a few other key contributors (see appendix C). A summary of the INTOR conceptual design was published in the IAEA journal *Nuclear Fusion* (vol. 22, p. 135, 1982).

The original INTOR conceptual design is depicted in figure 3.4, which shows a cross section of the toroidal configuration. The toroidal plasma is indicated by the empty teardrop shape to the right and left of the center, beneath which the poloidal divertor is located, as shown on the left of the drawing. Surrounding the plasma is a toroidal first-wall chamber (hatched), behind which are first an annular toroidal blanket and then a shield. The large boxes to the right and left are two of the six ion sources for the six-port neutral beam injection system used for plasma heating. One of twelve D-shaped toroidal field magnets is indicated on the right. The

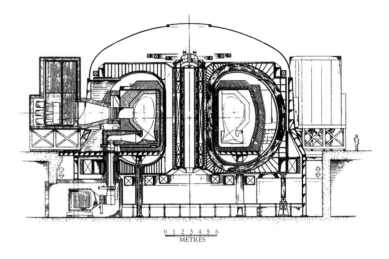

Figure 3.4   INTOR conceptual design (1981). See text for details.

poloidal field coil system consists of the central solenoidal magnet surrounding the central flux core, and the ring coils indicated at the top and at the bottom by the squares with an "X" in them (see figure 3.1 for a simpler schematic of the magnet systems). One of several large cryogenic pumps for maintaining the vacuum within the plasma chamber is shown at the lower left.

USA, Summer 1981

The TMB met July 28, 1981, in Los Angeles. The FED concept was turning out to be similar to the EC's JET soon to become operational, except the FED had superconducting coils. There was considerably more technical discussion than in previous TMB meetings, involving the same issues that had been discussed for INTOR at the same stage a couple of years previously.

The emerging concept was that FED could achieve similar physics objectives to INTOR, and then some other unspecified device could achieve the INTOR engineering testing objectives. The results from FED and the unspecified engineering device, taken together, would then allow the building of a demonstration fusion reactor that would perform well enough to inspire the confidence of utility executives. This is an appealing philosophy, if it is not examined too closely.

The appeal of the emerging FED concept, particularly to physicists, was that all the new technology that was needed to extract energy from fusion and all the engineering testing that was necessary before building a demonstration fusion reactor could be deferred to a later device (i.e., forgotten for now), and the present FED device could be dedicated to achieving ignition—the self-sustained plasma energy balance in which fusion heating was adequate to compensate radiation and transport losses. This "physics device plus engineering device" concept was not new; one variant or another rationalized the different physics "ignition devices" that had been put forward before by researchers in Europe, Japan, the USA, and the USSR who wanted to complete the physics development for fusion before confronting the engineering challenges of a fusion reactor.

This concept had already been examined in several venues, including most recently a year ago in the INTOR Workshop, where it had been concluded that the physics and engineering challenges of a fusion reactor are interrelated and must be addressed in an integrated fashion in an EPR. Some of us in the room had been through these arguments before, but the majority did not really want to hear them.

At the conclusion of the meeting, John Clarke of the DoE summarized recent developments in the DoE with respect to FED. The fusion program office had received internal DoE instructions to eliminate the FED initiative and reduce the budget accordingly, but since the official who had issued these instructions had left the government the next day, there was some hope that these instructions might be reversed (or forgotten). Clarke went on to outline the DoE's plans to have the fusion community work on critical technical issues affecting the success of FED while industry was carrying out competing designs of FED during 1982.

* * *

At this point (summer 1981), the INTOR Workshop had been by and large successful in developing an international consensus, at the technical level, on the feasibility, physical parameters and required R&D of an EPR. The problem now was to build support at the governmental level for moving ahead with the necessary design and R&D project. The first priority of the U.S. fusion program leader, Ed Kintner, was to build a national U. S. experimental power reactor (by whatever name), although it was becoming apparent that this was not going to happen under the new U.S. government. The fledgling European NET effort on a EPR was just getting started, and there was certainly no strong technical nor government support for moving forward with either a European EPR or INTOR at the moment. There was some support for INTOR, but certainly not a commitment, in the Japanese government. Only the USSR was firmly committed at the technical and government levels to moving forward into the design and construction of INTOR.

4

Phase 2A of the INTOR
Workshop (1981–88)

Although the fusion scientists and engineers working together in the INTOR Workshop were convinced by the summer of 1981 that it was technically feasible to move ahead with the design, procurement, and construction of a tokamak experimental fusion power reactor, the members of the IAEA's International Fusion Research Council (IFRC), consisting of the leaders of the government fusion programs of the INTOR Parties (USA, USSR, EC, Japan) and of other countries with fusion research programs, were not all prepared, for various reasons, to recommend that their governments move forward to a more detailed design activity eventually involving a centralized design team.

The first priority of the U.S. government fusion program leader, Ed Kintner, was developing support within the U.S. government and Congress to build a similar facility as a U.S. national project. He took the position on INTOR that the governments must identify a feasible administrative arrangement for carrying out an international construction project before undertaking a detailed design.

The USSR government fusion program leader, Yevgeny Velikhov, was strongly in favor of moving INTOR ahead into the design phase of an international project. The USSR government had already imposed constraints on the priority accorded to fusion research within the Soviet system, and prospects for the USSR building a device such as INTOR as a national project were undoubtedly perceived as quite small by those with any knowledge of the true state of the Soviet economy, Velikhov probably among them.

The various European fusion program leaders, including the EC fusion program leader, Donato Palumbo, were only beginning to seriously consider such a major "next-step" facility and were not ready to make any decision one way or another about it being constructed by the EC, much less internationally. In any case, a major review of the EC fusion program was imminent, which clearly precluded any EC decision to participate in a central INTOR design team before the completion of that review.

Japan was undergoing a financial crisis, which would have made a decision to participate in a central design team difficult, but not impossible, to reach. There was an appreciation at the technical level of the benefit of collaboration with the USA on such an undertaking, and there was some support in the Japanese government for such an international facility. It is likely that the Japanese would have supported a central INTOR design team if the other Parties, particularly the USA, had done so.

USA, August 1981

The INTOR Steering Committee made a presentation of the INTOR conceptual design to the IFRC at their meeting in Washington, D.C., on August 6, 1981. The Steering Committee also provided its cost estimate ($800 million for the device, $1.5 billion total direct cost, $3 billion total cost). The IFRC members had the following reactions (which are reproduced without editing from my journal):

1. Bol (Netherlands): Skeptical. Spell out complementary test program.
2. Palumbo (EC): Reasonable. Would like to see NET [Next European Torus] merge with a more international project. Look to reduce cost.
3. Von Gierke (Germany): Design is probably what any group would come up with for the same objectives.

4. Kakihana (Japan): Japanese government feels it should be constructed internationally. The world needs such a machine.

5. Kintner (USA): Reduce cost; implications for future unacceptable.

6. Pease (UK): Not surprised at cost. Press on.

The IFRC members then shared opinions about the continuation of the INTOR Workshop.

1. Bol (Netherlands): Consider a normal coil option.

2. Kintner (USA): FED [U.S. DoE's Fusion Engineering Device] input will be the U.S. contribution.

3. Palumbo (EC): What would be the cost if each party provided one-quarter of device?

4. Pease (UK): Include the tasks discussed by the IFRC subgroup.

5. Velikhov (USSR): Supports moving ahead to a design phase.

The continuation of the INTOR Workshop into Phase 2A through 1982, with three sessions in the remaining months of 1981 and three or four sessions in 1982 was then approved. A cost–benefit analysis was requested by June 1982, and a report analyzing the critical scientific and engineering issues confronting the success of INTOR was requested by December 1982. The IAEA was requested to maintain arrangements for the INTOR Workshop until June 1983.

The Terms of Reference for Phase 2A of the INTOR Workshop, as formulated by IFRC Chairman Bas Pease (UK) and transmitted to the IAEA director general, were as follows:

i) To perform a cost-benefit-risk analysis in which a variation of the objectives and parameters of INTOR is carried out to see the effect on costs, risks and time schedule both of the INTOR device itself and of the fusion development programme.

ii) To examine the potential impact of foreseeable advances in physics, such as steady current operation and radio-frequency heating.

iii) To analyze in greater depth certain critical issues which profoundly affect the design, such as: mechanical configuration and maintainability; tritium breeding and permeation; first wall design, impurity control and divertor configurations.

iv) To outline the design of advanced testing facilities of the torus, such as a combined tritium breeding and electricity generation segment, and of complementary non-fusion test facilities.

v) To define specific research and development projects required for the design of INTOR.

vi) To produce a report on the optimization of the conceptual design.

Then there was an exchange of views on the formation of an inter-governmental committee on administrative arrangements for a central team design Phase 2B of INTOR:

1. Pease (UK): No action is safest course.

2. Kintner (USA): No major design effort will take place until there is an indication it will lead somewhere.

3. Kakihana (Japan): The official response is not as positive as is the present position of the government.

4. Velikhov (USSR): The USSR is prepared to move ahead. It would be helpful to have a brainstorming session.

It was agreed that the IFRC subgroup (chairman plus representatives of USA, USSR, Japan, and EC) would meet November 16, 1981, in Brussels. Each member would bring two experts and written proposals regarding the possible administrative arrangements for the design phase.

* * *

The Steering Committee (Boris Kadomtsev, Gunter Grieger, Ken Tomabechi standing in for Sigeru Mori, and myself) flew to Atlanta for a meeting at Georgia Tech the following Monday, August 10. On Sunday afternoon I took them to Stone Mountain, an enormous granite outcropping with the icons of the Southern Confederacy (Robert E. Lee, Stonewall Jackson, and Jefferson Davis) astride their horses carved on the side. On the old-fashion train ride around the base of the mountain, they were hugely amused by a mock train robbery staged by guys on horseback that looked like they could have been in that business in the old days. Then we cruised on a paddle-wheel steamboat, were shown around transplanted plantation houses and slave quarters by pantalooned ladies with bustles and real Southern drawls, and went to my house for drinks (not mint juleps, unfortunately).

On the way to dinner, I drove the Russians (the Russian IAEA scientific secretary, Vladimir Vlasenkov, was with Kadomtsev) down Habersham Road through one of Atlanta's grander residential areas, describing mansions as typical homes (I probably even slipped in something about "worker's homes"), while my wife followed with Grieger and Tomabechi.

I took them all into the Lenox Square shopping mall but could not get the Russians beyond the Circuit City shop at the entrance, where we left them with instructions not to wander. We found them later standing outside laden with electronics. Then we all had a dinner of crayfish jambalaya and shrimp etouffee in Joe Dale's Cajun restaurant a few blocks down Peachtree Street.

The next day we met in my conference room at Georgia Tech for Session I of Phase 2A of the INTOR Workshop, where we planned the Phase 2A INTOR Workshop activities, which were quite different from the continuation and intensification of the INTOR design that we had been anticipating.

*   *   *

On August 21, 1981, I attended another U.S. Technical Management Board (TMB) meeting, this one at the Lawrence Livermore National Laboratory in California. Present were John Clarke and Lenora Ledman (DoE), Bob Borchers and Keith Thomassen (Livermore), John Rawls and John Gilleland (General Atomics), Harold Furth and Paul Rutherford (Princeton), John Sheffield (Oak Ridge), Charlie Baker (Argonne), Chuck Flanagan (FEDC), Bruce Montgomery (MIT), and myself. Flanagan told us about the cost and schedule for the Fusion Engineering Device (FED, the U.S. counterpart device to INTOR), and Borchers told us about a Livermore idea for a facility for engineering testing based on the tandem mirror plasma confinement concept dear to Livermore. The day ended with an inconclusive discussion of an organization of the FED activity involving competing industrial design teams and the fusion labs and universities. This discussion was picked up again the next morning, but without any further resolution.

*   *   *

August was clearly the month for meetings that year. An organizational meeting for INTOR Phase 2A was held at Georgia Tech on August 27, 1981. The attendees were John Schmidt and Paul Rutherford (Princeton), John Rawls (General Atomics), Bruce Montgomery (MIT), Mohamed Abdou (Argonne), Tom Shannon (FEDC), and myself. A major objective was to determine what part of the work that was planned for the FED could also be used for INTOR. We reviewed the INTOR and FED plasma design parameters and concluded that the ranges of anticipated operating physics parameters for the two designs overlapped sufficiently that any studies for one would also pertain to the other. We reviewed the list of critical technical issues that had been identified for FED studies and the proposed FED critical issues groups for possible use in the INTOR Workshop.

Vienna, September 1981

Session II of Phase 2A of the INTOR Workshop was held in Vienna at the IAEA International Conference Center on September 7–11, 1981. Assistant Director General Mauricio Zifferro welcomed us on behalf of the IAEA and praised the INTOR Workshop for "being ahead of the governments and causing them to think hard about their fusion programs." He then noted the two main tasks of the Phase 2A INTOR Workshop that had been recommended by the IFRC: (1) the analysis of critical technical issues that might affect the successful performance of INTOR, and (2) a cost/risk/benefit analysis of various capabilities in INTOR.

Zifferro went on to tell us that the IFRC had asked for an extension of the INTOR Workshop through June 1983, but that the IAEA was not prepared to take the leading role in contacting governments in this regard. He further told us that the IAEA understood that action at the governmental level was required for the advancement of the INTOR Workshop toward the realization of the actual device, and that the IAEA would take some action in this regard.

The participation in the Phase 2A INTOR Workshop had been expanded relative to earlier phases. The Participants were as follows: EC—G. Grieger (Germany), G. Casini (Italy), F. Engelmann (Netherlands), F. Farfaletti-Casali (Italy), M. F. A. Harrison (UK), A. F. Knobloch (Germany), D. Leger (France), P. Reynolds (UK), P. Schiller (Italy), and R. Verbeek (Brussels; scientific secretary); Japan—S. Mori, N. Fujisawa, T. Hiraoka. H. Iida, S. Nishio, and K. Tomabechi (JAERI), T. Honda, Y. Sawada, and T. Uchida (Toshiba), T. Kobayashi and T. Suzuki (Kawasaki), and K. Miyamoto (University of Tokyo); USA—W. M. Stacey (Georgia Tech), M. A. Abdou (Argonne), D. B. Montgomery and R. J. Thome (MIT), J. M. Rawls (General Atomics), J. A. Schmidt (Princeton) and T. E. Shannon, S. S. Kalsi, and T. G. Brown (FEDC); USSR—B. B. Kadomtsev, B. N. Kolbasov, A. S. Kukushkin, V. I. Pistunovich, and G. E. Shatalov (Kurchatov Institute), G. F. Churakov, A. I. Kostenko, S. N. Sadakov, and D. V. Serebrennikov (Efremov Institute), and V. G. Vasil'ev (Inorganic Materials Institute). Not all Participants attended all sessions.

The INTOR Workshop Phase 2A was initially organized into nine topical groups, with chairmen chosen to equitably distribute responsibilities among the Parties insofar as practical: (A) plasma performance (Engelmann, EC), (B) impurity control and first wall (Schmidt, USA), (C) testing (Tomabechi, Japan), (D) tritium (Abdou, USA), (E) mechanical configuration (Shannon, USA), (F) magnets (Knobloch, EC), (G) safety (Kolbasov, USSR), (H) cost and schedule (Hiraoka, Japan), and (I) cost/risk/benefit analysis (Steering Committee).

The tasks of groups A–G were to analyze critical technical issues that affected the success of the INTOR design, assess the impact of anticipated advances in the underlying physics and engineering technology on the design, identify possible improvements that could be made in the design, and update the Zero Phase R&D assessment. Group H was tasked to evaluate the cost and schedule of construction and operation of INTOR. Group I was tasked to evaluate cost, risk, and benefit of various INTOR operational and testing capabilities. These were important tasks that needed to be done, but the members of the INTOR Workshop were all disappointed that we were not proceeding with primary emphasis on further development of the INTOR design.

The Steering Committee members had a long discussion about the stalemate within the IFRC regarding the INTOR Workshop and the inability of the participating governments, or at least the mid-level representatives of those governments at the fusion program director level that constituted the IFRC, to agree to move forward to design and construction. The impediment was not based on technical concerns, and not even on high-level government or political constraints, but was rather largely a lack of willingness on the part of these middle-level bureaucrats either to push the matter to a higher level in their governments or to find a common ground for moving forward until that was possible.

My impression from those discussions, from my observations of IFRC meetings, and from my knowledge of the U.S. fusion program is the following. The government of the USSR was prepared to move forward to design and construction on the basis of the work of the first two phases of the INTOR Workshop. The Japanese government was moving slowly toward the same position and would have been responsive to strong American leadership to join together to design and construct INTOR. The EC government would not have been quite ready at that time, because they were rather later than the others in focusing on technology and fusion reactors, but they probably would not have wanted to be left out either.

The problem was the lack of positive leadership on the part of the USA. Ed Kintner, the U.S. Department of Energy (DoE) fusion program director, was definitely interested in moving forward to the design and construction of a major next-step tokamak device, but he favored the USA moving ahead unilaterally, perhaps with other nations participating in a U.S. project. Kintner pointed out the need to determine, at a high governmental level, if there was a likelihood for creating an administrative arrangement for moving forward to an international construction project before going into a detailed design phase. This made sense, of course, but it pushed the European and Japanese IFRC members harder than they felt they could go at the moment, and Kintner was unwilling to compromise. This position, of course, complemented his strategy for the USA to move forward unilaterally to build this major next-step device as a national project, the FED, rather than as an international project. However, as it was turning out, the new administration under President Ronald Reagan would not be supportive at a higher governmental level of any such expensive national fusion initiative.

The Steering Committee members felt a responsibility to keep the INTOR Workshop going in order to maintain active international cooperation in the development of the major next-step fusion device, in the event that some change in circumstances would come about to bring the governments together to find a way to move the activity forward into the design phase. We agreed that the work that we were doing needed to be done before the detailed design of such a device and would have been done in Phase 2A of the workshop in any case. All four of us were convinced of the logic of doing this step—developing an experimental fusion power reactor based on the workshop decisions already made—internationally and were resolved to do what we could to influence the situation favorably in our respective governments. I'm afraid in my case this was not much because both the director of the U.S. DoE fusion program, Ed Kintner, and his deputy and heir apparent, John Clarke, were determined to build the U.S. FED as a national project.

USA, Fall 1981

We were all quite busy in the short period between workshop sessions pulling together the material identified in the INTOR homework tasks. The U.S. INTOR Participants (Abdou, Schmidt, Shannon, Rawls, and myself) met in Atlanta on October 26, 1981, and again with Argonne and Fusion Engineering Design Center (FEDC) members of the U.S. INTOR team in Chicago on November 24, 1981, to review and discuss ongoing work on the INTOR homework tasks.

Vienna, December 1981

Session III of Phase 2A of the INTOR Workshop was held in Vienna on December 7–18, 1981. Returning to a two-week session allowed the workshop to return to its established rhythm of group meetings, plenary sessions, Steering Committee meetings, national team meetings, and discussions over coffee twice daily and frequently over wine and dinner in the evening.

Since the INTOR conceptual design provided an established set of reference parameters for all the critical issue studies, the coordination of studies was much easier than in previous workshop sessions, and the focus was on the technical details. The work now was more like what might take place within a research laboratory where several

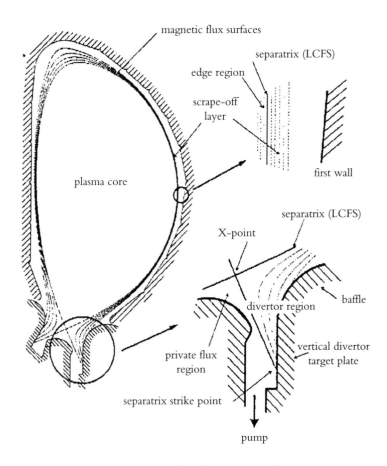

Figure 4.1   Lower single-null poloidal divertor configuration. See text for details.

High Density Divertor

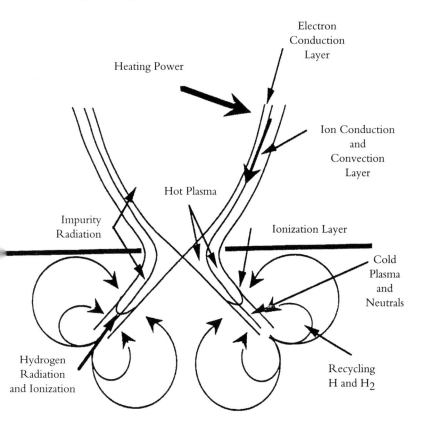

Figure 4.2 Processes occurring in a divertor.

people were working as a team. In fact, a large number of people in the four home teams were working as about a dozen teams, with their principal members meeting together in Vienna to compare results and plan new lines of inquiry. It was the function of the INTOR Workshop to focus those studies by making decisions among options on the basis of their effect on the INTOR design and defining home tasks for the periods between workshop sessions.

As one example, the details of the poloidal divertor concept, with high recycling of plasma particles depicted in figures 4.1 and 4.2

(which has subsequently been confirmed on several tokamak experiments), were worked out in the INTOR Workshop in this phase. The basic idea of the divertor is that energetic particles (ions and electrons) escaping the plasma across the last closed (magnetic) flux surface (LCFS) in the plasma chamber and entering the "scrape-off layer" are diverted along magnetic field lines into a separate divertor chamber where they are incident on a target plate.

These energetic incident ions "sputter" impurity atoms from the target plate and also recycle as neutral atoms of the plasma fuel (deuterium and tritium). There is a better chance that the sputtered "impurity" atoms can be prevented from entering the plasma, and that the recycling to the plasma chamber of the fuel atoms reflected from the target plate can be controlled, if this interaction with the material wall is in a divertor chamber removed from the main plasma chamber than if it occurred in the plasma chamber, which would be the case without a divertor.

In order to quantify this concept, mathematical models were developed for the various physical processes indicated in figure 4.2 and used to calculate the heat and particle fluxes escaping from the plasma that were incident on the "target" plates and to calculate the rate at which atoms were sputtered from these plates and transported back into the plasma core to increase the radiation cooling. Vladimir Pistunovich (USSR), John Schmidt (USA), Mike Harrison (UK), Andrei Kukushkin (USSR), later Doug Post (USA), and their colleagues in the INTOR home teams played key roles in working out these processes and their consequences in the INTOR Workshop.

Another workshop study of an entirely different character was the cost/risk/benefit study of various operational and testing capabilities in the INTOR design. This study was carried out by identifying and quantifying the physics and engineering "database" (i.e., the understanding and achieved performance parameters of the underlying physics and engineering) needed for the design of the demonstration reactor, or DEMO, that would follow INTOR. The importance of each piece of information needed for a DEMO design was quantified. Then the potential that an alternative (other than in INTOR) way to obtain that information was quantified, and so on, until finally a quantitative measure of importance for obtaining a specific piece of information in INTOR was determined. Then the incremental cost to the INTOR design of providing this piece of information in INTOR was calculated. Finally, a quantitative measure of the risk that the DEMO would not achieve its objectives in the absence of that piece of information was evaluated.

All of these quantitative measures were then combined to obtain a figure of merit for obtaining that information in INTOR. For example, one study (1) identified different threshold levels of neutron fluence (neutron flux × time of irradiation) required to obtain different levels of information about materials damage, component failure, and so forth; (2) evaluated the incremental cost of designing INTOR to produce the different levels of neutron fluence; (3) determined the cost and possibility of obtaining the different levels of neutron fluence for materials testing in facilities other than INTOR; and then (4) evaluated the risk of proceeding to construct the follow-on DEMO without having that information.

* * *

Historically (i.e., because one or more previous Participants had done so) the U.S. INTOR Workshop Participants lodged in one of three *pensions* in the central First District of Vienna inside the Ringstrasse. I stayed in Pension Riemergasse, just a few blocks from the Stephansdom, and the other U.S. team members stayed within a few blocks. The Japanese Participants stayed nearby in the Hotel Elizabeth near the old Fleischmarkt, while the Europeans favored a pension outside the Ringstrasse, and the Soviets stayed in dormitories owned by the USSR near the IAEA International Conference Center on the Danube. We would frequently come across each other on the weekends or in the evenings strolling in the First District.

At the end of the working day, the U.S. and Japanese team members took the U-Bahn from the IAEA International Conference Center to the central station at the Stephansdom and emerged onto the intersection of the two major pedestrian streets in the First District, the Kärtnerstrasse and the Graben. At this time of year the streets were lit by lampposts between which were hung green boughs with entwined white lights. This pre-Christmas season of 1981 was particularly fine, with snowflakes frequently in the crisp air and music everywhere in the evenings. The numerous churches held musical performances ranging from Mozart to Christmas carols that could be found just by strolling down the streets after work.

There was also the usual fine selection of concerts that Christmas season. I enjoyed a choral concert by the Vienna Bach Society in the Minorettenkirke on our first Sunday, a *Klavierabend* of Beethoven and Schumann at the MusikVerein on Tuesday, a Beethoven concert by the Wiener Symphonica at the Weiner Konzerthaus on the next Sunday, and the *Barber of Seville* at the Staatsoper during the second week.

The U.S. team had by this time established a tradition of a *heuriger* evening in one of the wine restaurants in the outlying villages that lay at the foot of the hillside vineyards surrounding Vienna. The new wine from that year's harvest was served by the pitcher (you could actually watch the sediment settle) and accompanied by such delicacies as *schmaltz* (chicken fat with bacon bits) and gorgonzola spread with garlic, both of which were to be spread on thick brown bread. Baked chicken, bratwurst, potato salad and other more familiar fare was also available for the faint of stomach. We usually went to Figlmüller's or the old Der Rudolfshof in Grinzing, taking the number 38 strassenbahn from the Schottentor to its end in Grinzing. However, this session we joined some Austrian friends at Killerman's in Perchtoldsdorf.

USA, Winter 1982

The situation of the FED (the U.S. counterpart to INTOR) within the U.S. fusion program was changing. Ed Kintner had retired as head of the DoE Office of Fusion Energy program, and John Clarke had taken his place. Don Steiner had resigned from Oak Ridge to become a nuclear engineering professor at Rensselaer Polytechnic Institute, and Tom Shannon had become the director of the FEDC with responsibility now for the FED work as well as the INTOR work at the FEDC. Tom agreed to continue his responsibility for the INTOR critical issue study on mechanical configuration but designated Paul Sager as the FEDC contact for work on the INTOR cost/risk/benefit study.

At a TMB meeting on January 19, 1982, at DoE Headquarters in Germantown, Maryland, John Clarke summarized the testimony of DoE director of the Office of Energy Research, Alvin Trivelpiece, to the House Science and Technology Committee. Trivelpiece was a plasma physicist who was now in a position of at least some influence in the government, but his message of the new administration's position was discouraging for the DoE fusion energy program.

Trivelpiece had testified to the House committee that the role of the government was to develop the database (basic scientific and engineering understanding) for fusion, but the demonstration of fusion as an energy source was the role of industry. This is a standard position of Republican administrations, but this administration was drawing the line between development and demonstration at an earlier stage. What this meant immediately was a very limited design effort on FED (ironically, the same restriction that Ed Kintner had contrived to impose on INTOR so that there might be more support within the government for a design effort on FED).

Trivelpiece had gone on to tell the Congress that in light of the constrained budgetary situation, the USA should be more serious about international cooperation, and that he was going to Japan and Germany to urge them to undertake important fusion technology development activities that the USA could not afford. Clarke informed us that he was trying to arrange an international cooperation between the FED project and a small Japanese design study of a similar device called FER (Fusion Energy Research [Facility]).

I found it ironic and a bit demoralizing that this new turn toward international collaboration did not seem to include the existing INTOR collaboration, which had defined an international consensus of the R&D that was needed and was an existing vehicle for initiating collaborative R&D projects. Since Trivelpiece was a plasma physicist capable of understanding all aspects of the INTOR work, the only explanation I could think of was that he was just not being told anything about the INTOR collaboration by Clarke (or previously Kintner), who reported to him in the DoE hierarchy.

Trivelpiece had also testified that the centerpiece of the Magnetic Fusion Energy Engineering Act of 1980, the Center for Magnetic Fusion Engineering, would disappear as a line item in the 1983 budget.

Clarke reported that Tom Shannon had replaced John Gilleland as executive director of the TMB, that DoE would follow up on the establishment of a committee on administrative arrangements for

INTOR, and that he (Clarke) intended to propose a new INTOR-type activity on the tandem mirror concept at the next meeting of the IFRC, which made me cringe. (The U.S. fusion program was developing the tandem mirror confinement concept at Lawrence Livermore National Laboratory, but it was far less advanced than the tokamak, and in fact the U.S. development of the tandem mirror would be terminated in 1986.)

This turn of events once again made INTOR the official principal focus in the U.S. fusion program for the examination of technical issues related to the major next-step tokamak experimental power reactor. The recent INTOR Steering Committee agreement to try to keep INTOR going in the event that circumstances changed seemed at this point to have been justified, even though the direction of the ongoing change in the USA was certainly not clear.

*   *   *

A workshop was held at Georgia Tech on March 8–10, 1982, to discuss and review the INTOR homework tasks on critical technical issues for the INTOR conceptual design in preparation for the March INTOR session. We had established a review committee consisting of several senior members of the U.S. fusion program (Charlie Baker, Argonne; Ron Davidson, MIT; John Gilleland, General Atomics; Paul Haubenreich and John Sheffield, Oak Ridge; Carl Henning, Livermore; Jerry Kulcinski, Wisconsin; Don Kummer, McDonnell Douglas; and Paul Rutherford and Paul Reardon, Princeton). The committee agreed that the right critical issue tasks had been identified and that we had the right people working on them. They recommended a major technical review before the June INTOR session.

Vienna, March 1982

Session IV of Phase 2A of the INTOR Workshop was held March 22 through April 2, 1982, at the IAEA International Conference Center in Vienna. Figure 4.3 shows a photograph taken during a plenary session of the workshop.

All four Parties had done a substantial amount of work in response to the homework tasks, and the workshop members were hard-pressed to digest and compare the results during the two weeks.

Figure 4.3 A plenary session of the INTOR Workshop Phase 2A, Session IV, Vienna, March, 1982: S. Mori and W. Stacey at head of table; to the right, G. Grieger, F. Engelmann, T. Hiraoka, and K. Tomabechi; to the left, counterclockwise, A. Knobloch, B. Kadomtsev, G. Shatalov, Y. Sawada, A. Kostenko, ?; against wall to left, G. Churakov.

The poloidal divertor for impurity control was looking better each session, and analyses of heating the plasma with electromagnetic waves instead of neutral beams were promising, particularly with waves at the ion cyclotron resonance frequency (ICRF) of about 80 million cycles per second, which is in the radiofrequency range for which highly developed power source technology existed. The investigation of the possibility of reducing the toroidal field coil size seemed to be confirmatory, but there was still difficulty in identifying a satisfactory poloidal field coil system that could produce the magnetic configuration needed for the poloidal divertor. The analysis of required blanket testing capability confirmed the need for a neutron fluence of about two megawatt-years per square meter, somewhat lower than had been previously thought. The calculations of tritium permeation into the first wall and the divertor target plates were still differing among the national teams by a factor of ten, which was attributed to differences in the simulation models and input data being used. Based on the progress by the end of the session, the Steering Committee determined that three more sessions, in July, October, and January, would suffice to complete and document the work of Phase 2A.

The Steering Committee discussed developments in the national fusion programs. Sigeru Mori stated that there was not yet an official Japanese position on entering the INTOR design phase, Phase 2B, but that more people in government were becoming interested in international cooperation. He also mentioned that the Japanese Fusion Council had approved the FER (the Japanese version of a primarily physics next-step option like the previous U.S. FED) but that it was his opinion that Japan would not build it alone because of cost.

Gunter Grieger reported that voices were growing louder in Europe for building their own European device, particularly among the fission reactor people who were now getting involved. He also stated that the administrative committee for INTOR had triggered a complaint among the European fusion laboratory directors that the IFRC had overstepped its authority, and that the new European science minister had brought the question of INTOR Phase 2B to the European Council of Ministers.

(Little Boris) Kolbasov, standing in for (Big Boris) Kadomtsev, reported that the USSR position was unchanged—they were ready to go into Phase 2B.

I reported the developments in the U.S. program mentioned above.

*   *   *

The IAEA director general had a center box at the Wiener Staatsoper, and it was possible to get seats for the opera when he was not using it. Viennese love their opera; the Staatsoper was the first building that they chose to have restored after World War II. I can't say that I became an opera buff, but I enjoyed going, almost as much for the opulence of the house and the scene at intermission as for the performances. Viennese dressed to the hilt, with a flute of champagne in one hand and the ever-present cigarette in the other, promenaded through the gilded lounges at intermission. I saw Mozart's *Abduction of the Seraglio* this session, and many others during my Vienna interlude.

The jewel of Vienna was to my mind the Weiner Philharmoniker, one of the world's truly great orchestras, performing in the

MusikVerein, a marvelous gilded and chandeliered hall much beloved by the Viennese. The few available public tickets were sold out weeks before a performance. By this time I had come to know a few Austrian friends who would purchase tickets for me for those few performances that coincided with INTOR Workshop sessions. This was a treat that I enjoyed only a precious few times during these INTOR years.

*   *   *

Even though the personnel changed, the U.S. INTOR Workshop delegation continued to be a close-knit group. Several traditions developed over the years. One of them was the final U.S. INTOR team meeting held in my office after the final plenary session, where we exchanged views about how things had gone and generally unwound and enjoyed each other's company for a few unhurried moments.

USA, Spring 1982

Upon returning to the USA, I attended a TMB (the Technical Management Board that loosely oversaw both the INTOR and current U. S. national EPR design efforts) meeting at DoE in Germantown, Maryland, on April 16, 1982. Other attendees were John Clarke (now director of the DoE fusion program upon Ed Kintner's retirement), Tom Shannon (now director of the FEDC at Oak Ridge), Keith Thomassen (Livermore), Bob Conn (UCLA), John Rawls (General Atomics), John Sheffield (Oak Ridge), Bruce Montgomery (MIT), Paul Rutherford (Princeton), and Charles Head (DoE).

Clarke presented me with a framed print of a *New Yorker* cartoon (figure 4.4) that had independently struck us both as particularly apt to our situation.

Budgets and their effect on future directions of FED were the main topic of the meeting. Clarke stated that we needed a strategy to justify the $500 million per year fusion budget that by default had become much larger than other decreasing program budgets in DoE. The problem, in his view, was finding a strategy that met current policy constraints against building a reactor (i.e., the FED). Clarke felt that a budget at that level could not be justified by "just doing

"I blame everything on the Russians, except for the few things, Stacey, I blame on you!"

Figure 4.4   Print of a *New Yorker* cartoon presented to the author by John Clarke caricaturing a meeting of the U.S. TMB. (© The New Yorker Collection 1984 Dana Fradon from cartoonbank.com. All rights reserved.)

research." He told us that the U.S. fusion program needed to maintain a mission orientation toward a "best design" for a FED.

* * *

Since the FED and INTOR critical issues analyses were being carried out in parallel, with a significant overlap in content and people, I attended the FED/INTOR Workshop on Ion Cyclotron Resonance Heating on May 6, 1982, and the review of the FED designs on May 10–11, 1982, at Oak Ridge. The "FED-R" design, headed by Dan Jassby of Princeton, was a copper magnet design similar to the existing European JET except for the added shielding that gave it the ability to operate with a deuterium plus tritium (D+T) plasma, hence a fusion plasma, while "FED-A" was the designation for the super-conducting magnet device more like INTOR. There was a good technical discussion of driving the plasma current with electro-magnetic waves rather than with inductive transformer action by Miklos Porkolab (MIT), Rip Perkins (Princeton), and others, which

I subsequently made arrangements to have included in the INTOR material being prepared for the next session.

* * *

The people working on the INTOR cost/risk/benefit study (Kulcinski, Wisconsin; Baker, Argonne; Steiner, Rensselaer; Clarence Trachsel, McDonnell Douglas; Sager, FEDC; and myself) met May 19, 1982, at O'Hare airport in Chicago.

A coordination meeting of the FED/INTOR managers; as we were now known (Schmidt, Princeton; Shannon, FEDC; Rawls, General Atomics; Abdou, Argonne; Dick Thome, MIT; and myself) was held at Princeton on May 27, 1982.

A large FED/INTOR meeting was held at Georgia Tech on June 29, 1982, to review the material prepared for the upcoming INTOR session in Vienna. This meeting was attended by the U.S. INTOR Participants (Stacey, Schmidt, Shannon, Rawls, Abdou, Thome), the people involved in doing the analyses, and the FED/INTOR review committee (Baker, Davidson, Gilleland, Haubenreich, Sheffield, Henning, Kulcinski, Kummer, Rutherford, and Reardon).

Several encouraging results were discussed. It was reported that recent experimental results from the DIII-D tokamak at General Atomics seemed to demonstrate that the poloidal divertor was more suitable than a pumped limiter for achieving desirable radiative cooling of the plasma edge, adding yet another reason for choosing this divertor option for impurity control. Calculations indicated that electromagnetic wave current drive could work for INTOR, although the computer model for the ICRF (ion cyclotron resonance) heating was not yet ready, so we would not have those results for Vienna. The analysis of testing capability confirmed the previous INTOR finding that there was a strong incentive to have a neutron fluence of two to three megawatt-years per square meter. The review committee was satisfied with the results, so we made only minor revisions and were off for Vienna.

Vienna, July 1982

Session V of the Phase 2A INTOR Workshop was held at the IAEA International Conference Center in Vienna on July 12–23, 1982.

The U.S. attendees were Tom Brown and Tom Shannon (FEDC), Mohamed Abdou and Dale Smith (Argonne), John Rawls (General Atomics), John Schmidt (Princeton), Dick Thome (MIT), and myself. The other attendees are listed in appendix B. The technical results tended by and large to confirm the directions in which the critical issues studies had been going at the previous session. The workshop tentatively agreed to reduce the size of the toroidal field coils, resulting in a large cost reduction, based on the U.S. analysis and subject to confirmation by the other three teams as a homework task for the next session.

The Steering Committee and the leaders of the critical issues groups, meeting together as a coordinating committee, found that enough conclusions on improvements to the design would be reached to make it feasible to define an improved design concept in the October session. A straw vote of the workshop showed a unanimous wish to make use of the information we had developed over the past year to improve the INTOR conceptual design. The Steering Committee members were aware of the strong opposition to further design work by John Clarke, the U.S. IFRC member (Clarke had now replaced Kintner, but took the same position), but we thought it was allowable under the IFRC Terms of Reference for Phase 2A of the INTOR Workshop because of item (iv), "To produce a report on the optimization of the conceptual design" (see beginning of this chapter). However, we refrained from defining design tasks that would have ensured the self-consistency of this new concept in the interest of not causing difficulty in the IFRC.

In a private discussion with Sigeru Mori, I learned that the Japanese FER device, the equivalent of the U.S. FED, was unlikely to be funded because of budgetary constraints.

*   *   *

It was the USA's turn to host an INTOR Workshop dinner, which we did in the spectacular surroundings of the Hotel im Palais Schwarzenburg. We enjoyed wine, campari and soda, and hors d'oeuvres on the terrace overlooking the mile-long vista along the Belvedere Park to the Palace of the Upper Belvedere housing the Klimts and Schieles. Then dinner was served for three dozen or so in an elegant dining salon from the Hapsburg era. It was customary on such occasions for the host Steering Committee member to make a few welcoming and motivational remarks. The circumstances, the setting, and the fact that it was the eve of my forty-fifth birthday prompted me to give a rather longer toast than is my norm in welcome of our guests, the handwritten copy of which I found stuck in my journal:

Gentleman and Madam,

I welcome you on behalf of the American INTOR Participants. It is our pleasure to have you join us this evening.

All of us here share a common dream of seeing fusion developed for the practical benefit of our fellow man.

We, in our INTOR Workshop, have had a unique opportunity to contribute to the realization of this dream.

At this time, we can look back and say that the INTOR Workshop has been a marvelous success. This success was due in part to the technical excellence of the people involved, in the home countries and here in Vienna. However, more than technical excellence was required. The success of the INTOR Workshop was due also to our being able to rise above personal ego and instinctive feelings of national pride to create a spirit of common purpose and fraternity.

INTOR has been a resounding success in technically advancing fusion. Our INTOR Workshop also has been an outstanding example that collaboration in the pursuit of a common objective benefits all.

Gentlemen and Madam, I propose a toast to the spirit of INTOR which unites us here in common cause.

(The one woman present was Doriana Twerksy, the Russian editor of the IAEA journal *Nuclear Fusion*.)

USA, Fall 1982

We returned home to make final calculations and prepare our final input to the INTOR Workshop Phase 2A, which would be published as a "national INTOR report," as was now the custom of the INTOR Workshop. The other Parties were doing the same.

The U.S. fusion program had also been active over the summer. DoE had once again changed the name of the U.S. major next-step device from the Fusion Engineering Device (FED) to the Engineering Test Reactor (ETR), perhaps to avoid prohibition from higher levels of government to continue work on FED, or perhaps because there is a natural cycle for this renaming obsession, as had been suggested in a humorous and widely circulated memo by Dan Jassby of Princeton.

Charles Head, the DoE program manager now responsible for the ETR and INTOR, called on September 21, 1982, to inform me that the DoE funding decision for near-term reactor studies had been made for 1983. There would be no funds for continuation of the critical issues studies for ETR, hence none for INTOR, after October 1. Charles instructed me that I should inform the FED/INTOR managers that there would be no funds to complete the ongoing studies. He also told me that John Clarke wanted me to inform the INTOR Workshop that the USA would not be doing critical issues studies (the very studies insisted on by the U.S. IFRC member a year earlier) during fiscal year 1983 (beginning October 1, 1982), but that we could inform the INTOR Workshop about what we were doing on studies of upgrading existing experiments. He went on to inform me that Clarke wanted me to get the INTOR Workshop to develop an option for Phase 2B of the INTOR Workshop for presentation to the IFRC that had as few sessions as possible in the first half of 1983, with those sessions being devoted to exchanging information about whatever research was ongoing in the four programs.

I reminded Head that the U.S. critical issues studies for FED and INTOR were in the process of being completed at the moment and would be published in a U.S. report in October and used as input to the October INTOR Workshop session, where a draft international report of Phase 2A would be prepared for review and finalization at the December session. So, except for documenting it, the work on critical issues had already been completed, and I strongly urged that the U.S. INTOR Participants be allowed to take part in completing the international INTOR report.

I also told Head that I would convey what he had just told me to the INTOR Workshop as "my instructions from my government" if that was what Clarke really wanted, but that I advised reconsideration because it would be strongly resented by our international partners as an attempt to trivialize the INTOR Workshop. In any case, I told him, it was the IFRC that would decide on the Terms of Reference for the next phase of the workshop, and Clarke, as a member, could express his views directly to them. Head said that he would get back to me on this, but he never did, and I interpreted this as a nonconfirmation of his instructions to deliver this message to the INTOR Workshop, justifying my decision not to do so.

* * *

A FED-INTOR meeting was held at Georgia Tech on September 28, 1982, to review the material to be taken to the October INTOR session in Vienna. The next day, September 29, 1982, I received a phone call from Manfred Leiser, the IAEA (USA) scientific secretary for INTOR in Vienna, informing me of a new problem for the INTOR Workshop from an unexpected source. At an IAEA Board of Governors meeting the previous day, Iraq had sponsored a resolution to expel Israel, the USA had threatened to walk out, and the motion had failed. Then Iraq had sponsored a resolution to withdraw credentials from Israel, which failed to carry on a 40–40 tie vote. An absent delegate from Madagascar was later found strolling about Vienna, brought back to the board meeting, the vote was reopened, and the resolution passed 41–40, upon which the U.S. delegation had walked out.

An immediate halt of all participation of DoE-sponsored people in IAEA activities had been ordered by the DoE. According to Leiser, the U.S. State Department was formulating a policy, but the likelihood of the USA participating in the October INTOR Workshop session seemed small to him, and his own status was uncertain. Charles Head (DoE) called on October 4, 1982, to inform me that there would be no U.S. participation in the October INTOR session. I called Sigeru Mori in Tokyo, and we decided to postpone the October session until this blew over.

* * *

In October 1982, the U.S. FED-INTOR report was published and distributed in two volumes in the now trademark blue binders. The list of contributors was six pages long, with representatives from Argonne National Laboratory, Burns and Roe, EG&G Idaho, Exxon Research and Engineering, FEDC at Oak Ridge, General

Atomics, Grumman Aerospace, General Electric, Georgia Institute of Technology, Hanford Engineering Development Laboratory, Los Alamos National Laboratory, Lawrence Berkeley Laboratory, Lawrence Livermore National Laboratory, McDonnell Douglas, MIT, Monsanto's Mound Laboratory, Oak Ridge National Laboratory, Pacific Northwest Laboratory, Princeton Plasma Physics Laboratory, Pennsylvania State University, Rensselaer Polytechnic Institute, Sandia National Laboratory, University of California, University of Illinois, University of Michigan, University of Wisconsin, and Westinghouse.

My cover letter summarizes the contents of the report:

This is your copy of the FED-INTOR report, in two volumes. The report documents the work that was carried out in the FED-INTOR activity during 1981–82 with the objective of advancing our understanding of the technical and programmatic issues affecting the concept for an experimental tokamak reactor that could follow the upcoming generation of large tokamaks (TFTR, JET, JT60, T-15). This work involved a large segment of the U.S. fusion community and was conducted in concert with the INTOR Workshop: as such, this report represents the U.S. contribution to that Workshop.

Several technical issues that affect the feasibility, cost and engineering tractability were studied in detail: bulk heating by ICRF, impurity control, tritium containment, mechanical configuration and maintenance, magnets and electromagnetics. Comprehensive reviews and assessments were performed in several areas: plasma confinement and beta limits, plasma-wall interaction, tritium permeation, etc. The engineering testing requirements were examined in great detail, and a cost-risk-benefit study was performed to provide perspective on mission definition.

Based upon the conclusions and recommendations that were derived from these studies, the FED and INTOR designs which had been developed in 1981 were evolved toward an improved design concept. This improved concept represents our recommendation for the starting point in defining further conceptual design work for a tokamak experimental reactor, ETR in the USA and INTOR internationally.

A summary paper was published in the journal *Nuclear Technology/Fusion* ("The FED-INTOR," vol. 4, p. 202, 1983).

* * *

IAEA's 9th biennial International Conference on Plasma Physics and Controlled Nuclear Fusion Research was held in Baltimore in late fall of 1982, with a plenary session on INTOR. Sigeru Mori (Japan) introduced the session, I gave the overview talk on the INTOR Workshop Phase 2A results, Boris Kadomtsev (USSR) gave the physics summary talk, John Schmidt (USA) gave a talk on the poloidal divertor, Tom Shannon (USA) gave a talk on the mechanical configuration, Al Knobloch (EC) gave a talk on the magnet systems, and Gunter Grieger (EC) gave a talk on tritium.

Vienna, April 1983

Session VI of Phase 2A of the INTOR Workshop was finally held on April 11–22, 1983, at the IAEA International Conference Center in Vienna. Each of the four Parties brought its national INTOR report as input. The content of these reports had been discussed and iterated at previous INTOR Workshop sessions, so the main task of this session was to reach final decisions on a few items and to merge these national reports into a single INTOR report that could be left with the IAEA to produce page proofs for review at the session now planned for August.

The Steering Committee (Mori, Kadomtsev, Grieger, and myself) discussed developments relative to INTOR in our governments.

Mori reported that the completion of JT60, which had experienced significant delays and cost overruns, was the first priority in the Japanese fusion program, and that both JAERI and the government were reluctant to discuss the proposal for the FER (a device equivalent to the U.S. FED) because of the economic depression in Japan. There was still a commitment to continue INTOR, but not for too much longer in the present state.

Kadomtsev reported that completing T-15 remained the highest priority for the USSR fusion program. He stated that they were willing to continue INTOR for two more years in the present state, but then a decision must be made to go forward with a project or to stop.

Grieger reported that JET was a high priority in Europe. He stated that a Next European Torus (NET) group had been formed under Romano Toschi to coordinate work on the next-step device and technology, that in the future INTOR input would be provided completely by the NET team, and that three NET division leaders would join INTOR in the next phase. He reported that the mood in Europe was that it was still too soon to agree to produce an engineering design of INTOR until JET produced significant results in about 1986.

I summarized the situation in the U.S. fusion program discussed above in the best light possible.

The Steering Committee then discussed future plans for the INTOR Workshop. Once again, we were unanimous that we should try to find a way to keep the workshop going as a vehicle for maintaining momentum and international cooperation on the next-step device, while governments searched for mechanisms to enable design and construction to go forward. We were well aware by now, of course, that the problem was not lack of mechanisms as much as the lack of will to find them on the part of some of the middle-level bureaucrats who ran the government fusion programs.

Continuation of the critical issues studies, refinement of the R&D assessment, evolution of the design concept and similar topics were agreed to as reasonable suggestions for the next phase of the INTOR Workshop, but we all realized that these were holding actions.

On April 4, 1983, the Steering Committee met with the IFRC. Mori described the status of the INTOR Workshop Phase 2A and the preparation of the report, which should be finalized at the current session and approved for publication at the next session in August. I described our suggestions for activities in the next phase. The IFRC instructed us to add to the list of items to be investigated in the next phase (1) safety and decommissioning, (2) disruption control, and (3) definition of the benefits of different levels of physics and technology testing. All IFRC members then endorsed a two-year extension of the INTOR Workshop with the proposed activities.

* * *

Vienna is famous, among other things, for its humorous operettas, perhaps the best known of which is *Die Fledermaus*. This session I attended a performance of this kitschy Viennese favorite in the traditional hall for operettas, the Theater an der Wien. I thoroughly enjoyed the experience of the performance and the audience reaction, although most of the humor was lost on me because of my limited ability to understand colloquial Viennese.

Vienna, August 1983

Session VII of the INTOR Workshop Phase 2A was held at the IAEA International Conference Center in Vienna on August 1–5, 1983. The first few days were devoted to proofreading the Phase 2A INTOR Workshop report, which was subsequently published by the IAEA as STI/PUB/638. This report had 150–200 contributors from each of the four Parties and was authored principally by the INTOR Workshop Participants, with the support of the IAEA Scientific Secretariat M. Leiser (USA) and V. S. Vlasenkov and A. A. Shurygin (USSR). The contents and authors of the report are given in appendix C of this book.

The report is succinctly summarized in the forward by Hans Blix, director general of the IAEA:

The Phase-Two A, Part I Report of the International Tokamak Reactor (INTOR) Workshop is the third in a series of reports documenting the activities of the world's leading fusion countries in the conceptual design of the next-generation tokamak experiment. Phase 2A, Part 1, has emphasized the resolution of certain critical technical issues identified in Phase One which affect the feasibility, cost and engineering complexity of the INTOR design concept. This work continues into Part 2 of Phase Two A. The INTOR Workshop finds it importance not only in defining the technical issues facing the construction of such a machine, but also in providing an important example of international collaboration in a field where successive generations of experiments are beginning to tax the resources of all but the richest countries.

As noted in the director general's introduction, the IFRC had side-stepped the issue of moving forward to Phase 2B, which had been defined as a design phase with a recommendation for the formation of a permanent central team. Instead, they had redefined the current Phase 2A to have two parts, Part I which had just been completed, and part II which was to come.

A summary of Phase 2A (Part I) of the INTOR Workshop was published in the IAEA journal *Nuclear Fusion* (vol. 23, p. 1513, 1983).

Vienna, January 1984

Session VIII of Phase 2A of the INTOR Workshop met at the IAEA International Conference Center in Vienna on January 16–27, 1984. The Participants for this session are given in Appendix B.

For the new Phase 2A Part II, the workshop was organized into eight groups: (1) impurity control (physics head Doug Post, USA, and engineering head Peter Schiller, EC), (2) radiofrequency heating (head Folker Engelmann, EC), (3) transient electromagnetics (head Al Knobloch, EC), (4) maintenance (head Ken Tomabechi, Japan), (5) technical benefit (head Ken Tomabechi, Japan), (6) physics (head Vladimir Pistunovich, USSR), (7) nuclear (head Gely Shatalov, USSR), and (8) engineering (head Chuck Flanagan, USA). These groups met to plan their work for the next two years of Phase 2A Part II and to define specific homework tasks for the next session.

There was the traditional INTOR dinner, hosted this time by the Japanese delegation, in a dark, candlelit Viennese restaurant to welcome the new Participants and solidify friendships among the old-timers. In addition, I further introduced the new U.S. team to the charms of Vienna with an evening at the famous Figlmüller *heuriger* in the old wine village of Grinzing on the outskirts of Vienna.

London, September 1984

The results of the first part of Phase 2A were presented in an INTOR plenary session at IAEA's 10th biennial International Conference on Plasma Physics and Controlled Nuclear Fusion Research in London in September 1984. The Steering Committee reported to the IFRC at their meeting in the Royal Society headquarters in London (figure 4.5).

The London conference was particularly memorable for me on two counts. At a dinner for the INTOR Steering Committee and the IFRC at the Royal Society headquarters I learned that I was sitting where Newton once sat when he presided over Royal Society meetings.

Even more eventful for me was meeting my second wife in a South Kensington restaurant the next evening. She had gone with her brother to celebrate selling a large painting to Scott's, another London restaurant. Bob Conn, John Gilleland, and I walked into a crowded restaurant after a reception at the National Museum of

Figure 4.5   A break in the IFRC meeting, Royal Society, London, September 1984. Left to right: Donato Palumbo, Bas Pease, Bill Stacey.

Science and Industry and ended up sharing their table. Before the evening was out I had bought a painting she had on exhibition in yet another restaurant. I was more interested in the artist than the painting but had been praising the latter when John Gilleland suggested (loudly enough to be heard by all) that if I liked it so much I should buy it, which of course I did. A few years later I got the artist, too, without any push from John this time.

Summary of Phase 2A Part II (1984–85)

The INTOR Workshop met five times during Part 2A Part II at the IAEA International Conference Center in Vienna: Session VIII, January 16–27, 1984; Session IX, May 21 through June 1, 1984; Session X, October 15–27, 1984; Session XI, April 14 through May 3, 1985; and Session XII, September 9–14, 1985. Homework tasks were carried out by large teams in the home countries between sessions, as in the past. The Participants and expert attendees for Phase 2A Part II of the INTOR Workshop are given in Appendix B.

As before, the INTOR work was carried out by about 150 scientists and engineers working in the fusion laboratories, industries, and universities of each of the four Parties, under the guidance of the INTOR Participants. The work was reviewed in detail at each INTOR session, and homework tasks were defined for performance in the home institutions before the next session.

In the USA, in addition to numerous telephone calls to arrange and coordinate the allocation of effort to perform the INTOR homework tasks, nine coordination or review meetings were held involving the people doing the work, the U.S. INTOR Participants, and the ten people mentioned previously who were serving as the U.S. INTOR review committee. The work of each party was compiled into a national INTOR report that was used to prepare the final international INTOR report. The U.S. national INTOR report was published in two volumes in the now traditional blue binders, and similar reports were published by the other three Parties. A summary of the U.S. contribution to this phase of INTOR was published in the journal *Fusion Technology* (vol. 11, p. 317, 1987).

Advances in the understanding of several issues important to the INTOR concept were achieved during Phase 2A Part II:

1. the physics of heating and driving current in tokamak plasmas with electromagnetic waves in the radiofrequency range;

2. the physics of the poloidal divertor;

3. the engineering of first wall and divertor target concepts that could withstand the interaction with the escaping energetic plasma ions and electrons;

4. improved concepts for more compact and more robust superconducting magnets;

5. improved poloidal field coil configurations for forming and maintaining the poloidal divertor magnetic configuration and for controlling the plasma, now taking into consideration the various complicated paths for current flow in the structures of the device; and

6. a mechanical configuration that would allow for all-remote maintainability.

The INTOR design concept was updated to reflect these advances in understanding but, in compliance with the IFRC guidance not to perform a new design (based on the insistence of John Clarke [USA] that no design work be carried out), a new self-consistent design was not developed.

The Steering Committee members were hopeful that the operation of the new generation of large tokamaks (TFTR, JET, JT60, T-15) would in the near future provide the governments with the incentive to move forward with the INTOR design phase. However, we were aware of the plethora of proposals being made for less advanced national physics experiments in the event that INTOR could not move forward.

In consultation with the IFRC, the Steering Committee prepared recommendations for a next phase of the INTOR Workshop. We wanted to continue exploration of the above issues and their effect on the design, to evaluate the impact of new results from the tokamak physics experiments on the INTOR design, and to develop a revised self-consistent conceptual design based on the developments of the past few years. The IFRC suggested that it also would be useful to make a critical comparison of the various new proposals for national next-step tokamak experiments with the INTOR concept. Also, the IAEA in the normal course of business held Specialist Committee meetings on various fusion-related topics that were subject to IFRC approval, and the IFRC asked that these meetings be held under the aegis of the INTOR Workshop to provide them with a focus.

* * *

We were in Vienna during ball season a couple of times during Phase 2A. On one occasion, I managed to get a ticket for one of the grandest, the Philharmonia Ball, held in the spectacular gilded halls of the MusikVerein. The orchestra hall, converted to a dance hall, has frescoed ceilings three stories above hung with glittering chandeliers. I found an excellent vantage point up a set of broad marble stairs on a balcony overlooking the dance floor. The festivities started at 9 P.M. with several dozen young men and women from one of the Viennese dancing schools swirling elegantly about the floor, exhibiting the waltz as it was meant to be done. Then came an orchestral fanfare and the formal opening of the ball by the president of Austria, the lord mayor of Vienna, and a host of lesser dignitaries and their ladies, followed after the first dance by hundreds of elegantly attired Viennese sweeping around the dance floor. This went on until dawn, with pauses for champagne, gravlax sprinkled with capers, and other delicacies, followed by the traditional afterball repast of *profiteroles* (cream puffs) and black coffee in a nearby restaurant.

On another occasion, the entire U.S. delegation was invited to a private ball in the apartment of one of our Austrian friends, where we spent a very enjoyable, if somewhat less elegant, evening.

* * *

The INTOR Workshop report published by the IAEA as STI/ PUB/714 was authored by the INTOR Participants, with 500–600 contributors. The foreword to this report by Hans Blix, director general of the IAEA, very well summarizes its purpose and content:

> The Phase Two A, Part II report of the International
> Tokamak Reactor (INTOR) Workshop is the fourth in a

series of reports documenting the activities in the world's leading fusion countries in the conceptual design of the next generation tokamak experiment.

Phase Two A, Part II has emphasized the resolution of certain critical issues essential to the feasibility and improvement of the INTOR concept. The physics and engineering data bases which support the INTOR design concept have also been re-evaluated.

Based upon the new information resulting from the critical issues studies and the data base assessment and upon the additional design scoping studies carried out by the four Participants, the INTOR design concept was slightly modified. The modified INTOR design is, in part, based on extrapolations of the data available at present which could be substantiated by the current generation of large tokamak experiments (TFTR, JET, JT60, T-15) before the project enters the detailed design phase.

The success of this work and the desirability of continuing this useful international activity led the International Fusion Research Council to recommend to the IAEA that it be continued for a further two and one half years.

Consistent with its objectives of providing a world consensus on the critical issues facing the construction of the next-generation large tokamak experiment, the INTOR Workshop will shift its emphasis during the next period to the consideration of possible innovations in the tokamak concept, and to the relationship of such a machine to the next step beyond, a Fusion Power Demonstration plant.

A summary of the material in this report was published in the IAEA journal *Nuclear Fusion* ("The Phase 2A, Part 2 INTOR Workshop," vol. 25, p. 1791, 1985), and a plenary session on INTOR was held at the IAEA's 11th biennial International Conference on Plasma Physics and Controlled Nuclear Fusion Research in Kyoto in the fall of 1986.

* * *

The Steering Committee continued to exchange information about the situation in the government fusion programs with regard to arrangements for moving INTOR into the design phase. No progress was apparent during Phase 2A, Part II. The USSR remained willing to move forward, Japan was having a financial depression and cost overruns on JT60, the USA was continuing to try to get a less ambitious national experiment funded, and the EC felt no pressure to come to any decision. The INTOR Workshop was coming to be considered by some plasma physicists and fusion engineers as just a paper study. Something different was needed to break the stalemate, and that something came along in the form of Mikhail Gorbachev.

Geneva Summit Meeting, November 1985

At his first summit meeting with U.S. President Ronald Reagan in Geneva in 1985, USSR Premier Mikhail Gorbachev proposed that the USA and USSR join together to design, construct, and operate a tokamak fusion energy reactor like INTOR. The hand of Yevgeny Velikhov in this proposal was unmistakable to all of us in INTOR, but none of us except undoubtedly Kadomtsev knew it was coming. It is unlikely that anyone on the U.S. side at the summit meeting knew what Gorbachev was talking about, but when they got home they quickly found Al Trivelpiece, director of the DoE's Office of Energy Research and a plasma physicist, who did. Al had a leading role in formulating the response, which brought in our Japanese and Europeans partners as well.

My impression from subsequent conversations with various Soviet colleagues in Vienna was that the U.S. bureaucracy almost managed to snatch failure from the jaws of success. The USSR state committee was insulted by the way the U.S. presidential initiative, as they termed the U.S. response, was handled. There apparently had been an agreement at Geneva for the USA and USSR to prepare papers for discussion of fusion cooperation that would be exchanged in December 1985. However, the response of the USA was instead an initiative to include the Europeans and Japanese and involve the IAEA, which came as an unannounced unilateral action. To exacerbate matters, in the Soviet view, the U.S. initiative was received by the IAEA on December 3, by the EC on December 4, and by the USSR only on December 6, 1985. Some in the USSR government felt that the U.S. initiative did not represent a sufficient commitment and might be a ploy to combine the USSR Geneva proposal and INTOR and kill them both with a paper study. Despite these suspicions, they went forward with the process.

When the pressure became top-down, rather than bottom-up, as it had been up to this point, it made all the difference. Within a year of Gorbachev's proposal there was an intergovernmental agreement to go forward with design and the supporting R&D program that had been identified by the INTOR Workshop. Al Trivelpiece and John Clarke, now a born-again internationalist, were at the center of the intergovernment negotiations that led to the formation of a project to move INTOR into the design phase. Of course, the natural law governing the name change cycle of fusion projects was still in force, and the project was renamed ITER, the International Thermonuclear Experimental Reactor, or in Latin "the way."

Summary of Phase 2A, Part III (1986–88)

Phase 2A Part III of the INTOR Workshop met at the IAEA International Conference Center in Vienna in four sessions: XIII, March 10–21, 1986; XIV, December 1–12, 1986; XV, July 13–24, 1987; and XVI, November 9–20, 1987. This phase of the workshop was planned before the Gorbachev initiative and changed only slightly as a result of the pending formation of the ITER project, which did not actually take place until 1988.

The Participants and experts attending the INTOR Workshop sessions were greatly expanded now that the prospects for design and construction were brighter. The EC INTOR Workshop attendees were as follows (see glossary for abbreviations): G. Grieger (Steering Committee, Germany), K. Borrass (NET), F. Casci (Italy), E. Cocceorese (NET), F. Engelmann (NET), F. Farfaletti-Casali (Italy), M. Harrison (UK), A. Knobloch (Germany), D. Leger (France), N. Mitchell (NET), E. Salpietro (NET), P. Schiller (Italy), W. Spears (NET), R. Verbeek (scientific secretary, Brussels), G. Vieder (NET), and J. Wegrowe (NET). The supporting home team consisted of about 150 scientists and engineers from EURATOM (JET, Oxford, and NET, Munich), Italy (ANSALDO, Genova; CNR, Milano; ENEA, Bologna and Frascati; JRC, Ispra; MATEC,

Milano; NIRA, Genova; and the universities of Cagliari, Calabria, Milano, Napoli, Salerno, and Trieste), France (CEA, Fontenay, Genoble, Saclay, and Cadarache; LAN, Paris; Technicatome, Saclay), Germany (HMI, Berlin; IPP, Garching; INTERATOM, Bergisch Gladbach; KfA, Julich; KfK, Karlsruhe; Siemens, Erlangen; universities of Braunschweig and Stuttgart), Netherlands (ECN, Petten; FOM, Niewegein; JRC, Petten), Belgium (CEC, Brussels; ERM/KMS, Brussels; CEN/SCK, Mol), UK (UKAEA, Harwell, Culham, Risely, and Warrington), Switzerland (CRPP, Lasaunne; SIN, Villingen), Sweden (Chalmers University; KTH, Stockholm; Studsvik Energiteknik, Studsvik), Denmark (DAE, Riso), Luxembourg (LUXATOM, Gradel), and Canada (University of Toronto).

The Japanese INTOR Workshop attendees were S. Mori (Steering Committee chair), N. Fujisawa, H. Iida, T. Kobayashi, K. Tomabechi, and T. Tsunematsu (JAERI), T. Honda (Toshiba), B. Ikeda, M. Kasai, and R. Saito (Mitsubishi), and T. Mizoguchi and T. Okazaki (Hitachi). The supporting home team consisted of about 150 scientists and engineers from Fuji Electric, Fujitsu, Kyoto and Nagoya universities, Hitachi, Kawasaki, Mitsubishi, Toshiba, and JAERI.

The U.S. INTOR Workshop attendees were W. M. Stacey (Steering Committee vice chair, Georgia Tech), D. A. Ehst and D. L. Smith (Argonne), C. A. Flanagan, M. Peng, and P. T. Spampinato (FEDC), and N. Pomphrey and D. E. Post (Princeton). The U.S. supporting home team consisted of about 150 scientists and engineers from Argonne, Canadian Fusion Fuels, DoE, EPRI, FEDC, Georgia Tech, Grumman, General Atomics, McDonnell Douglas, MIT, Princeton, Penn State University, the Sandia Laboratories at Albuquerque and Livermore, TRW, UCLA, Westinghouse Hanford, and the Idaho, Oak Ridge, Los Alamos, and Lawrence Livermore National Laboratories.

The USSR INTOR Workshop attendees were B. B. Kadomtsev, Y. L. Igitkhanov, B. N. Kolbasov, V. I. Kripunov, A. S. Kukushkin, V. I. Pistunovich, G. E. Shatalov, and V. L. Vdovin (Kurchatov Institute, Moscow), A. I. Kostenko, R. N. Litunovskij, and I. V. Mazul (Efremov Institute, Leningrad), S. A. Yakunin (Inorganic Materials Institute, Moscow), and A. M. Epinatiev (Energy Institute, Moscow). The supporting home team consisted of about 150 scientists and engineers from Moscow (Bajkov Metallurgy Institute, Keydish Mathematic Institute, Kurchatov Institute, Institute for Energy Technics, Inorganic Materials Institute, Institute of Physics and Chemistry, Moscow State University; Physical-Engineering Institute, and Lebedev Physics Institute), Kiev (Institute Electrodynamics and Institute of Cybernetics), Leningrad (Polytechnic Institute, Efremov Electrophysical Institute, and Ioffe Physico-Technical Institute), Kharkiv (Physico-Technical Institute), and Sukhumi (Physico-Technical Institute).

The INTOR Workshop followed the now familiar and well-functioning operational procedure of having the INTOR Participants and selected experts meet together in Vienna to review the work that had been performed in the fusion laboratories, universities, and industries in the home countries, to make decisions based on this review, and to define specific homework tasks to be performed before the next INTOR Workshop session. A notable new feature of Phase 2A Part III was that experimental information from the new generation of large tokamaks (TFTR, JET, JT60) was brought to the INTOR Workshop to be factored into evaluation of technical issues. There was some indication that the dependence of the energy confinement time (the time in which energy deposited in the plasma would be confined before leaking out) on plasma current was greater than previously thought. There were also substantial advances made in the understanding of (1) impurity control, (2) plasma operational limits, (3) electromagnetic wave current drive and heating, (4) electromagnetics, (5) configuration and maintenance, and (6) the engineering of the tritium breeding blanket and first wall facing the plasma were made.

The potential impact on the INTOR conceptual design of these various individual improvements in understanding was evaluated. In addition, a series of engineering scoping studies was performed to provide insight into the effect certain other design changes might have if they were incorporated into the design. The ways in which the INTOR conceptual design should be changed in view of the new understanding from this work were discussed and documented, but the actual updating of the INTOR conceptual design was left as a task for the design Phase 2B, which would finally start as the ITER activity in 1988.

There was some "chafing at the bit" in the INTOR Workshop during the latter part of Phase 2A Part III because of the desire to move ahead with an updated self-consistent conceptual design. (Many future members of the initial ITER design team were present.) However, the Steering Committee was aware of the delicate balance in the IFRC between Clarke's (USA) insistence that there should be no design work and the diverse but generally more positive opinions of the other members, which had resulted in the compromise IFRC instruction to update the design concept but not to develop an improved design. The Steering Committee agreed that the workshop should carefully adhere to this IFRC instruction rather than risk provoking a confrontation in the IFRC, given that the Gorbachev initiative seemed to be leading finally toward an agreement for detailed design and construction of INTOR.

The final INTOR design concept is depicted in figure 4.6. The most visible difference relative to the initial INTOR conceptual design of figure 3.4, at this level, is the replacement of the neutral beam heating system by an ion cyclotron resonance heating system and a lower hybrid resonance current drive system. The large boxes to the outside of the reactor vessel shown in figure 3.4 contained the ion sources for the neutral beam heating system, which are no longer present in figure 4.6. There were a multitude of other differences, as discussed above, but these affected the design concept at a less visible level. Certainly the new evidence of the stronger dependence of confinement time on plasma current would have resulted in an increased plasma current, but this would have required a redesign of the poloidal magnetic and torus support systems and undoubtedly would have led to a somewhat larger device.

* * *

A new aspect of Phase 2A Part III of the INTOR Workshop was the embedding of IAEA Specialist Committee meetings, bringing additional fusion scientists and engineers, many of whom were not otherwise directly involved in the INTOR process, together with INTOR specialists in several areas in order to focus deliberations on the impact of their findings on the INTOR conceptual design. Specialist Committee meetings were held on (1) impurity control modeling, (2) tokamak concept innovations, (3) demonstration

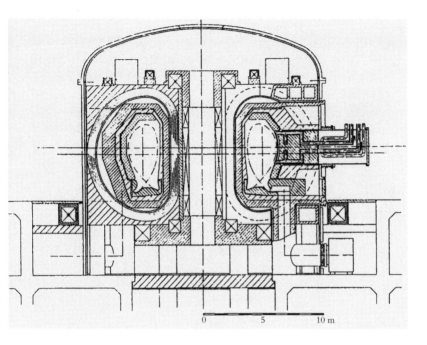

Figure 4.6   INTOR design concept, 1987. (The toroidal plasma is indicated by the symmetric teardrop-shaped dashed lines about 5 m from the centerline. The poloidal divertor chamber is below the plasma. Surrounding the plasma is a toroidal first wall, followed by a blanket and then a shield. Two of the twelve D-shaped toroidal field magnets are shown surrounding the torus. The poloidal field magnets are a central solenoidal magnet indicated on each side of the open central flux core and a set of toroidal ring coils indicated by the boxes with a cross.)

reactor requirements, (4) noninductive current drive, (5) confinement in tokamaks with intense heating, (6) plasma disruptions, and (7) comparison of INTOR-like designs.

The last item, comparison of INTOR-like designs, had been added by the IFRC after the Gorbachev initiative and involved bringing together the members of the teams that were designing proposed national tokamak projects—FER (Japan), NET (EC), OTR (USSR) and a new TIBER (USA) concept proposed by Livermore—with the INTOR Workshop members. The programmatic and technical objectives, physics and engineering constraints, design specifications, major parameters, materials, and so forth, were documented in a common format and compared with INTOR.

There was some resistance in the INTOR Workshop about spending time reviewing these national "INTOR-like" designs, which were far less thought through and developed than the INTOR design. Several of the members of the workshop, in particular the newer members, complained that this review of much less developed designs was a waste of valuable time. Ettore Salpietro, a vocal and energetic Italian member of the EC team, was particularly incensed by this task and let it be known at every opportunity.

However, the Steering Committee agreed that the workshop must carry out this task in good faith in order not to risk upsetting the delicate balance on the IFRC. We also appreciated the benefit of involving the proponents of the various designs in the INTOR process so that they would feel that their views had been taken into consideration, which would make them more likely to support the design and construction of ITER rather than holding out for their national project. Both Sigeru Mori and I made short talks in plenary sessions of the INTOR Workshop to encourage the workshop members to take this task seriously, and I held a number of private discussions when the opportunity arose. We simply did not want to rock the boat at this point.

*   *   *

The U.S. INTOR team held its final dinner for their INTOR colleagues in Vienna on December 8, 1986, and I offered what turned out to be my final INTOR toast, which was a bit on the long side, at least in part because of the need I felt to bring us back together to complete our task:

> Gentlemen, we are brought together once again in service of the INTOR Workshop.
> INTOR was born in a moment of high hopes, both for fusion and for international cooperation, created by the IAEA in response to the Soviet suggestion.
> The INTOR Workshop was as an experiment. We surprised everyone by showing that men of different cultures and strong opinions could work together, could put the common good of the workshop ahead of their personal pride,

and could voluntarily make decisions and abide by them to develop a single design concept.

INTOR succeeded in forging a worldwide agreement on the features of the next major step in the world tokamak program and on its necessity.

These initial successes of INTOR were followed by a time during which international cooperation was out of fashion in some governments. National design teams were formed, drawing upon INTOR for inspiration and for experienced people.

INTOR became a university of fusion during this period where people came together to learn and to advance their common knowledge and to develop solutions to technical problems.

Perhaps more important, INTOR continued to provide the world with an example of international cooperation during a period when this was unfashionable. INTOR kept alive the possibility of international collaboration on the next major tokamak.

Now, this possibility has been raised to the highest political levels. International cooperation is once again fashionable in the governments.

There are proposals from the USSR and from the USA for a joint ETR design project. It has been suggested that this project be hosted by the IAEA. The INTOR Workshop has been asked to take on tasks during 1987 aimed at laying the technical groundwork for such a project.

So, once again we are at a time of high hopes for international cooperation, and once again INTOR has an opportunity to demonstrate that such cooperation is possible.

We can look forward with confidence to building upon our previous work to lay the technical basis for an ETR design over this next year in our INTOR Workshop. Let us hope that the bureaucrats will be equally successful in reaching agreement on administrative matters.

I think that we all can take pride, as members of the INTOR Workshop, in a job well done under sometimes difficult circumstances, over these past eight years.

The possibility for realization of the common vision that has guided the INTOR Workshop—an international tokamak reactor—seems more promising now than at any time in the past.

I propose that we drink to that vision and to those members of the INTOR Workshop, past and present, who have risen to the challenge of making the INTOR process work so well over these eight years.

They probably were pretty thirsty after all that.

* * *

As before, the U.S. work in support of the INTOR homework tasks was performed by about 150 scientists and engineers working in the U.S. fusion laboratories, universities, and industry under the guidance of the U.S. INTOR Participants. In addition to the usual plethora of phone calls to arrange effort allocation and to coordinate tasks, and visits by the individual INTOR Participants to the sites where the work was taking place, seven major reviews of the U.S. work on the INTOR homework tasks were attended by the people performing the work, the U.S. INTOR Participants, and the U.S. INTOR review committee. Similar activities were carried out in Europe, Japan, and the USSR. This work was documented in national INTOR reports published in 1987 and brought to the final full session of the INTOR Workshop, Session XVI held November 9–20, 1987, as input for preparing the final INTOR report. A summary of the U.S. contributions to Phase 2A Part III of the INTOR Workshop was published in the journal *Fusion Technology* (vol. 15, p. 1485, 1989).

The brief foreword written by Hans Blix, director general of the IAEA, for the Phase 2A Part III INTOR Workshop report (STI/PUB/795, IAEA, Vienna) succinctly summarizes both this final report and the role of the INTOR Workshop in the development of fusion:

The Report of the International Tokamak Reactor Workshop for Phase Two A, Part III, is the fifth and final report in the series documenting the joint activities of the world's four major fusion blocks in the conceptual design of the next generation tokamak experiment, known under the acronym INTOR.

The activities of the Workshop during its last phase included a definition of the database for fusion, a study on possible innovations for a tokamak reactor and comparison of different national next generation tokamak concepts. The report will thus not only prove useful as the final documentation of eight years of concerted design work by the world fusion community, but will also be of great value in providing a foundation on which the work of INTOR's natural successor, the International Thermonuclear Experimental Reactor (ITER), can be based.

The preface by Max Brennan (Australia), then chairman of the IAEA IFRC, provides elaboration:

The work of the INTOR Workshop has been completed and this report—in two volumes—presents the results of the final phase of the Workshop—Phase 2A, Part III. This phase, conducted during 1985–1987, included a continuing assessment of the evolving tokamak physics and technology databases, an analysis of several critical technical issues and a number of possible innovations. The original work-plan for the last year (1987) also included an updating of the 1981 INTOR design concept to take account of the Workshop's studies of Critical Issues and the evolution of the database. With the decision of the European Community, Japan, the Soviet Union and the USA to embark on the ITER Project (International Thermonuclear Experimental Reactor), the work-plan was changed by the IFRC to focus on a critical analysis of the several existing INTOR-like designs.

In accepting the final report of the INTOR Steering Committee, the IFRC agreed with the Committee's conclusions that the extensive work of the INTOR Workshop had contributed to the technical understanding from which the ITER design could proceed toward its objective of a practical conceptual design. The Council also felt that the Workshop had done much of the exploratory technical work needed to proceed with the ITER design. The IFRC also agreed that the ability to work together on such a complex task, developed with considerable effort and skill by the INTOR Participants, would be of considerable benefit to the ITER activity.

In fact, the INTOR design and the information in the INTOR reports was carried forward directly as the starting point for further ITER design by the many INTOR Participants who formed the initial ITER team.

A summary of the final INTOR report was published in *Nuclear Fusion* (vol. 28, p. 711, 1988), and the detailed results were presented at a final INTOR plenary session at the IAEA 12th biennial International Conference on Plasma Physics and Controlled Nuclear Fusion Researchin Nice in October 1988.

The IAEA held a final reception for the INTOR Workshop, and the INTOR Steering Committee held its final meeting (figures 4.7 and 4.8) during this last full session of the INTOR Workshop in

Figure 4.7   Final INTOR Steering Committee meeting, Vienna, November 1987. Left to right: Norma ? (IAEA), Manfred Leiser (IAEA/USA), Sigeru Mori (Japan), Mauricio Zifferero (associate director general of IAEA), Bill Stacey (USA), Gunter Grieger (EC), Boris Kadomtsev (USSR), A. Shurygan (IAEA/USSR).

Figure 4.8 INTOR Steering Committee, Vienna, November 1987. Left to right: Bill Stacey (USA), Gunter Grieger (EC), Sigeru Mori (Japan), Boris Kadomtsev (USSR).

November 1987. Mori presented us all with INTOR pins with the INTOR design concept emblazoned on it, which we are wearing in these final photographs.

* * *

On a cold March 14, 1988, Gunter Grieger, Vladimir Pistunovich, Noboru Fujisawa (filling in for Mori), and I gathered one last time at our offices in the IAEA International Conference Center in Vienna to go over the edited page proofs for the final INTOR report. Grieger and I, who were the only ones to attend every meeting of the INTOR Workshop from November 1978 to March 1988, thus became the first and the last of the INTORians. We all finished our work by midafternoon, shook hands, and said goodbye, and the INTOR Workshop was history.

5

Epilogue

We had previously unsuspected poetic talent amongst us in those closing days of INTOR, as evidenced by an anonymous epilogue left in the coffee room during the last days of the INTOR Workshop:

Epilogue to INTOR Phase 2A

2B or not 2B: that is the question:
Whether 'tis Nobel-er for the world to persevere
In the face of adversity,
Or to respond against a sea of instabilities
And by resigning end them? To stop: to sleep;
No more; and, by a sleep to say we end
The debate and the thousand natural conflicts
That success is heir to, 'tis a consummation
Easy to accept. To stop, to sleep;
To sleep: perchance to dream; aye, there's the rub;
For in that sleep of resignation what dreams may be recalled,
Of the oft' told promises of this mortal torus
Must give us pause. There's the reason
That gives research so long a life;
For who would bear the whips and scorns of time,
Indecision over size, the proud defence of maintenance,
The pangs of anomalous transport, the funding's delay,
The insolence of office, and the spurns
Of politicians for finite, unwieldy cost,
When he might take his leave of phase 2A
By logging off his CRAY.

Who would the vagaries bear,
Of multinational diplomacy,
But that the expectation of success after INTOR
The undiscovered ETR derived
From Artsimovich's dream, drives the will,
And makes us rather bear those ills we have
Than fly to a future both dark and cold?
Thus conscience does make heroes of us all.

*   *   *

At Sigeru Mori's request and the Steering Committee's concurrence, I wrote the technical summary of the INTOR Workshop, which was published in the journal *Progress in Nuclear Energy* (vol. 22, p. 119, 1988). Now, twenty years later, perhaps enough time has passed to put into perspective the broader history of the INTOR Workshop in its successful quest for the first fusion energy reactor, which is the purpose of this accounting.

*   *   *

INTOR's successor, the International Thermonuclear Experimental Reactor (ITER) project, officially began in 1988 as a collaboration among the same INTOR Parties—USSR, USA, European Community (EC), and Japan. ITER was hosted for its first phase (1988–92) at the Max-Planck-Institut für Plasmaphysik at Garching, near Munich, Germany.

Many members of the INTOR Workshop continued their work, becoming the experienced core of the ITER project. Ken Tomabechi (Japan) was the first director of ITER, and John Gilleland (USA) was the U.S. representative on the Technical Management Committee for the new project. Folker Engelmann (EC) and Gely Shatalov (USSR) brought eight years of INTOR experience, physics and nuclear, respectively, to their new duties in ITER. Doug Post (USA) and Andrei Kukushkin (USSR) continued their poloidal divertor development work, which they had started in the INTOR Workshop, for the ITER project. Yuri Igitkhanov (USSR) and Noboru Fujisawa and Masayoshi Sugihara (Japan) became key members of the physics team for ITER. Alexander Kostenko (USSR), T. Tsunematsu (Japan), Ettore Salpietro and Bill Spears (EC), and Chuck Flanagan (USA) became leading members of the engineering team for the new ITER project, and Charlie Baker (USA) was an important contributor to the nuclear team. Dick Thome (USA) and several other INTO-Rians joined the ITER project at a later stage.

Boris Kadomtsev (USSR) and Paul Rutherford (USA) served on the ITER Science and Technical Advisory Committee, and I chaired the ITER Steering Committee–US (ISCUS), during the early years of the ITER project.

The ITER Engineering Design Phase (1992–2001), involving the same parties (with the USSR replaced by Russia and the EC replaced by the European Union), was initially led by Paul-Henri Rebut and then by Robert Aymar, both of France. The engineering design, carried out at three electronically interconnected sites in Germany, California, and Japan, was completed in July 2001, and $650 million worth of supporting R&D was completed shortly thereafter. This design and the adequacy of the supporting R&D were extensively reviewed by national and international committees, all of whom came to the positive conclusion that the ITER design would achieve the ITER mission. This design was subsequently revised somewhat to reduce the size and the cost.

The process of agreeing on the site for ITER construction was a lengthy one. Canada first offered a site in Clarington in 2001, followed soon after by a Japanese proposal of a site at Rokkasho-Mura, a Spanish proposal of a site at Vandellos near Barcelona, and a French proposal of the Cadarache nuclear research site in Provence. Canada withdrew its proposal in 2003, and the European Union chose the Cadarache site as the single European site proposal. After extended negotiations, a "broader approach" was agreed to in 2005 in which ITER would be sited at Cadarache in France, Japan would provide 20% of the staff for the ITER Project, Europe would make a fifth of its share of procurements in Japan, the project director would be proposed by Japan, and Japan and Europe would work together on other fusion program elements (e.g., an accelerator-based Fusion Materials Irradiation Facility and related engineering validation and design activities) at an International Fusion Research Center in Rokkasho, Japan.

Several changes occurred in the participants to the ITER project over the course of its existence. The Russian Federation replaced the USSR when the latter collapsed, the U.S. withdrew from 1999 to

2003, the People's Republic of China and the Republic of Korea both joined the project in 2003, and India joined in 2005. At this point, the governments involved in ITER represent more than half of the world's population.

Following the siting decision, the project moved toward its construction phase, which began in 2009. Agreement was reached on the sharing of costs and the in-kind contributions to the project, the joint implementation agreement was signed by all participants, the international ITER Organization was established in Cadarache, Kaname Ikeda of Japan (a diplomat with a nuclear engineering background) was appointed director general of the project, Norbert Holtkamp of Germany (an accelerator builder with a physics background) was appointed project construction leader, and the ITER scientific and technical team of several hundred people began to come together at Cadarache. A final design review was completed in 2008, with several design improvements being made as a consequence. As of this writing, expectation is that ITER will begin operation in 2018, forty years after the first meeting of the INTOR Steering Committee on the Boltzmangasse in Vienna set the process in motion.

Appendix A
Sessions of the INTOR
Workshop

Zero Phase

Steering Committee	November 20–23, 1978
I	February 5–16, 1979
II	June 11–July 6, 1979
III	October 1–19, 1979
IV (Steering Committee)	December 17–19, 1979

Phase 1

I (Steering Committee)	January 16–18, 1980
II	March 24–28, 1980
III	June 16–27, 1980
IV	October 20–31, 1980
V	January 19–February 4, 1981
VI	March 30–April 10, 1981
VII	June 22–July 3, 1981

Phase 2A

Part I

I (Steering Committee)	August 10–11, 1981 (Atlanta)
II	September 7–11, 1981
III	December 7–18, 1981
IV	March 22–April 2, 1982
V	July 12–23, 1982

Appendix B
INTOR Workshop Participants
and Experts

Asterisks (*) indicate individuals who participated as experts.

Table App. 2 INTOR Workshop Participants and Experts

EC	JAPAN	US	USSR
Phase Zero (1978–79)			
G. Grieger	S. Mori	W. M. Stacey	B. B. Kadomtsev
F. Engelmann	T. Hiraoka	J. R. Gilleland	G. F. Churakov*
R. Hancox*	K. Sako	G. L. Kulcinski	B. N. Kolbasov
D. Leger	T. Tazimi	P. H. Rutherford	V. I. Pistunovich
P. Reynolds	H. Momota*	C. C. Baker*	G. F. Shatalov
J. Bohdansky*	Y. Shimomura*	V. A. Maroni*	
G. Casini*		D. M. Meade*	
J. Darvas*		J. R. Alcorn*	
Phase One (1980–81)			
G. Grieger	S. Mori	W. M. Stacey	B. B. Kadomtsev
G. Casini	N. Fujisawa	M. A. Abdou	G. F. Churakov
F. Engelmann	T. Hiraoka	T. G. Brown	B. N. Kolbasov
F. Farfalletti-Casali	K. Sako	C. A. Flanagan	V. I. Pistunovich
R. Hancox*	M. Sugihara	G. L. Kulcinski	D. V. Serebrennikov

A. Knobloch	K. Tomabechi	J. A. Schmidt	G. E. Shatalov
D. Leger	S. Itoh*	T. E. Shannon	V. G. Vasil'ev*
P. Reynolds	Y. Sawada*		V. A. Loktev*
P. Schiller	T. Kobayashi*		
	N. Miki*		

Phase 2A, Part I (1981–83)

G. Grieger	S. Mori	W. M. Stacey	B. B. Kadomtsev
G. Casini	N. Fujisawa	M. A. Abdou	G. F. Churakov
F. Engelmann	T. Hiraoka	R. J. Klemmer	B. N. Kolbasov
F. Farfalletti-Casali	T. Honda	J. M. Rawls	A. I. Kostenko
M. Harrison	H. Iida	P. H. Sager	V. I. Pistunovich
A. Knobloch	T. Kobayashi	J. A. Schmidt	D. V. Serebrennikov
D. Leger	K. Miyamoto	T. E. Shannon	G. E. Shatalov
P. Reynolds	S. Nishio	R. J. Thome	
P. Schiller	Y. Sawada		
	T. Suzuki		
	K. Tomabechi		
	T. Uchida		

(*Cont.*)

Table App. 2 (*Continued*)

EC	JAPAN	US	USSR
Phase 2A, Part II (1983–85)			
G. Grieger	S. Mori	W. M. Stacey	B. B. Kadomtsev
H. Chazalon	N. Fujisawa	C. C. Baker	B. N. Kolbasov
F. Engelmann	T. Honda	P. L. Colestock	A. I. Kostenko
F. Farfalletti-Casali	H. Iida	C. A. Flanagan	A. S. Kukushkin
M. Harrison	S. Itoh	R. F. Mattas	R. N. Litunovskij
A. Knobloch	M. Kasai	M. Y. Peng	V. I. Pistunovich
D. Leger	H. Kimura	R. J. Pillsbury	D. V. Serebrennikov
E. Salpietro	T. Kobayashi	D. E. Post	G. E. Shatalov
P. Schiller	K. Miyamoto	P. T. Spampinato	
G. Vieider	M. Seki	R. J. Thome	
	M. Sugihara		
	K. Tomabechi		
	T. Tone		
	K. Ueda		

Phase 2A, Part III (1985–88)

G. Grieger	S. Mori	W. M. Stacey	B. B. Kadomtsev
K. Borrass	N. Fujisawa	D. A. Ehst	A. M. Epinatiev
F. Casci	T. Honda	C. A. Flanagan	Y. Igitkhanov
F. Engelmann	H. Iida	Y. M. Peng	V. I. Khripunov
F. Farfaletti-Casali	B. Ikeda	N. Pomphrey	A. I. Kostenko
A. Knobloch	M. Kasai	D. E. Post	A. S. Kukushkin
D. Leger	T. Kobayashi	D. L. Smith	R. N. Litunovsk
N. Mitchell	T. Mizoguchii	P. T. Spampinato	I. V. Mazul
E. Salpietro	T. Okazaki	J. R. Tarrh	V. I. Pistunovich
P. Schiller	R. Saito		G. E. Shatalov
W. Spears	T. Sekiguchi		V. Vdovin
G. Vieider	K. Tomabechi		S. Yakunin
J. Wegrowe	T. Tsunematsu		

Appendix C
Reports of the INTOR
Workshop

The INTOR Workshop produced five official reports (based on five sets of national INTOR reports by the four parties produced as input to the workshop).

"International Tokamak Reactor, Zero Phase," STI/PUB/556, IAEA, Vienna, 1980 (650 pages).

Contents: (I) Introduction; (II) Summary, Conclusions, and Recommendations; (III) Database and R&D Needs Assessments, INTOR Suggested Parameters; (IV) Energy and Particle Confinement; (V) Impurity Control, Fueling and Exhaust; (VI) Heating; (VII) Stability Control; (VIII) Startup, Burn, and Shutdown; (IX) Magnetics; (X) Power Supply and Transfer; (XI) First Wall, Blanket, and Shield; (XII) Tritium, (XIII) Materials; (XIV) Systems Integration and Support Systems; (XV) Assembly and Remote Maintenance; (XVI) Radiation Shielding and Personnel Access; (XVII) Vacuum; (XVIII) Diagnostics, Data Acquisition and Control; (XIX) Safety and Environment; Appendices.

Authors: EC—G. Grieger, J. Bohdansky, G. Casini, J. Darvas, F. Engelmann, R. Hancox, D. Leger, P. Reynolds; Japan—S. Mori, T. Hiraoka, H. Momoto, K. Sako, Y. Shimomura, T. Tazimi; USA—W. M. Stacey, C. C. Baker, J. R. Gilleland, G. L. Kulcinski, V. A. Maroni, D. M. Meade, J. R. Purcell; USSR—B. B. Kadomtsev, G. E. Churakov, B. N. Kolbasov, V. I. Pistunovich, G. F. Shatalov; IAEA—F. N. Flakus.

"International Tokamak Reactor, Phase One," STI/PUB/619, IAEA, Vienna, 1982 (860 pages).

Contents: (I) Introduction; (II) Summary; (III) Physics Design Basis; (IV) Mechanical Configuration and Maintenance; (V) Magnetic and Electrical Systems; (VI) Heating and Fueling Systems; (VII) First Wall and Limiter Systems; (VIII) Divertor and Divertor Plates; (IX) Tritium Breeding Blanket; (X) Radiation Shielding; (XI) Vacuum and Tritium Systems; (XII) Diagnostics, Instrumentation, Data Acquisition and Control; (XIII) Layout of Facilities; (XIV) Machine Operation and Test Program; (XV) Reliability and Availability Assessment; (XVI) Safety and Environmental Impact; (XVII) Site Criteria; (XVIII) Research and Development; (XIX) Schedule; (XX) Design Specifications; Appendices.

Authors: EC—G. Grieger, G. Casini, F. Engelmann, F. Farfaletti-Casali, R. Hancox, A. Knobloch, D. Leger, P. Reynolds, P. Schiller; Japan—S. Mori, N. Fujisawa, T. Hiraoka, T. Kobayashi, N. Miki, M. Sugihara, K. Sako, Y. Sawada, K. Tomabechi; USA—W. M. Stacey, M. A. Abdou, J. Alcorn, T. G. Brown, J. G. Crocker, B. L. Hunter, G. L. Kulcinski, G. D. Morgan, J. A. Schmidt, D. L. Smith, C. A. Trachsel; USSR—B. B. Kadomtsev, G. E. Churakov, B. N. Kolbasov, V. I. Pistunovich, D. V. Serebrennikov, G. F. Shatalov, V. G. Vasil'ev; IAEA—F. N. Flakus.

"International Tokamak Reactor, Phase Two A Part I," STI/PUB/638, IAEA, Vienna, 1983 (772 pages).

Contents: (I) Introduction; (II) Summary, Conclusions and Recommendations, (III) INTOR Concept; (IV) Plasma Confinement and Control; (V) Radio-Frequency Heating and Current Drive; (VI) Impurity Control Physics; (VII) Impurity Control and First-Wall Engineering; (VIII) Tritium and Blanket; (IX) Magnets; (X) Electromagnetics; (XI) Mechanical Configuration; (XII) Engineering Testing; (XIII) Cost and Schedule; (XIV) Cost-Risk-Benefit; (XV) Research and Development; (XVI) Design Specifications; Appendices.

Authors: EC—G. Grieger, G. Casini, F. Engelmann, F. Farfaletti-Casali, P. Harbour, M. Harrison, A. Knobloch, D. Leger, P. Reynolds, P. Schiller; Japan—S. Mori, N. Fujisawa, T. Hiraoka, H. Iida, T. Kobayashi, K. Miyamoto, S. Nishio, Y. Sawada, T. Suzuki,

K. Tomabechi; USA—W. M. Stacey, M. A. Abdou, T. G. Brown, R. F. Mattas, D. E. Post, J. M. Rawls, M. Rogers, J. A. Schmidt, T. E. Shannon, R. J. Thome; USSR—B. B. Kadomtsev, G. E. Churakov, B. N. Kolbasov, A. I. Kostenko, A. S. Kukushkin, V. I. Pistunovich, S. N. Sadakov, D. V. Serebrennikov, G. F. Shatalov, V. G. Vasil'ev; IAEA—F. N. Flakus.

"International Tokamak Reactor, Phase Two A Part II," STI/PUB/714, IAEA, Vienna, 1986 (849 pages).
Contents: (I) Introduction; (II) Summary, (III) Impurity Control; (IV) Radio-Frequency Heating and Current Drive; (V) Transient Electromagnetics; (VI) Maintainability; (VII) Technical Benefit of Partitioning INTOR Component Design and Fabrication; (VIII) Physics; (IX) Engineering Database Assessment; (X) Nuclear; (XI) INTOR Concept Evolution; (XII) Design Concept; (XIII) Operation and Test Program; Appendices.
Authors: EC—G. Grieger, M. Chazalon, E. Cocceorese, F. Engelmann, F. Farfaletti-Casali, P. Harbour, M. Harrison, A. Knobloch, D. Leger, E. Salpietro, P. Schiller, G. Vieder; Japan—S. Mori, N. Fujisawa, T. Honda, H. Iida, S. Itoh, M. Kasai, H. Kimura, K. Miyamoto, S. Nishio, Y. Sawada, M. Seki, M. Sugihara, K. Tomabechi, T. Tone, K. Ueda; USA—W. M. Stacey, C. C. Baker, P. L. Colestock, C. A. Flanagan, R. F. Mattas, Y. M. Peng, R. D. Pillsbury, D. E. Post, D. L. Smith, P. T. Spampinato, J. Stevens, J. M. Tarrh, R. J. Thome; USSR—B. B. Kadomtsev, G. E. Churakov, B. N. Kolbasov, A. I. Kostenko, A. S. Kukushkin, R. N. Litunovskij, V. I. Pistunovich, S. N. Sadakov, D. V. Serebrennikov, G. F. Shatalov; IAEA—F. N. Flakus.

"International Tokamak Reactor, Phase Two A Part III," STI/PUB/795, IAEA, Vienna, 1988, Vol. 1 (653 pages), Vol. 2 (329 pages).
Volume 1 Contents: (I) Introduction; (II) Summary, (III) Impurity Control; (IV) Operational Limits and Confinement; (V) Current Drive and Heating; (VI) Electromagnetics; (VII) Configuration and Maintenance; (VIII) Blanket and First Wall; (IX) Additional Physics Issues; (X) Additional Engineering Issues; (XI) INTOR-Related Activities; (XII) Conclusions about the INTOR Design Concept; Appendices.

Volume 1 Authors: EC—G. Grieger, F. Casci, M. Chazalon, E. Cocorese, F. Engelmann, F. Farfaletti-Casali, M. Harrison, A. Knobloch, D. Leger, E. Salpietro, P. Schiller, G. Vieder, J. Wegrowe; Japan—S. Mori, N. Fujisawa, T. Honda, H. Iida, B. Ikeda, M. Kasai, T. Kobayashi, T. Mizoguchi, T. Okazaki, T. Tsunematsu; USA—W. M. Stacey, D. A. Ehst, C. A. Flanagan, Y. M. Peng, R. D. Pillsbury, N. Pomphrey , D. E. Post, D. L. Smith, P. T. Spampinato; USSR—B. B. Kadomtsev, A. M. Epinatiev, B. N. Kolbasov, A. I. Kostenko, V. I. Khripunov, A. S. Kukushkin, R. N. Litunovskij, I. V. Mazul, V. I. Pistunovich, G. F. Shatalov, V. L. Vdovin, S. A. Yakunin.

Volume 2 Contents: (XIII) Report of IAEA Specialist Meeting on INTOR-Like Designs.

Volume 2 Authors: EC—G. Grieger, E. Salpietro, R. Toschi, F. Casci, M. Chazalon, E. Cocceorese, P. Dinner, F. Engelmann, F. Farfaletti-Casali, M. Harrison, D. Leger, N. Mitchell, P. Schiller, R. Verbeek, G. Vieder, J. Wegrowe; Japan—R. Saito, K. Tomabechi, N. Fujisawa, H. Iida, T. Kobayashi, N. Miki, M. Sugihara, M. Yamada; USA—W. M. Stacey, B. G. Logan, D. E. Post, C. C. Baker, J. N. Doggett, C. A. Flanagan, C. D. Henning, F. W. Perkins, D. L. Smith, P. M. Stone; USSR—V. I. Pistunovich, G. E. Shatalov, A. M. Epinatiev, V. I. Khripunov, A. I. Kostenko, R. N. Litunovskij, S. N. Sadakov, S. Yakunin.

Appendix D
Tokamaks in the World

Since being introduced by the T-1 in the USSR in 1957, about 200 tokamaks worldwide have operated or are under construction. The table below lists a representative set of the major tokamaks. A more complete listing may be found at www.tokamak.info.

First Year	Device	R (m)	a (m)	B (T)	I (MA)	Divertor	NBI P(MW)	ICRH P(MW)	LHR P(MW)	ECR P(MW)
1957	T-1[1]	0.63	0.13	1.0	0.04					
1962	T-3[1]	1.0	0.12	2.5	0.06					
1963	TM-3[1]	0.4	0.08	4.0	0.11					
1968	LT-1[11]	0.4	0.10	1.0	0.04					
1970	ST[2]	1.09	0.14	4.4	0.13					
	T-4[1]	1.0	0.17	5.0	0.24					
1971	Ormak[2]	0.8	0.23	1.8	0.20		0.34			
	T6[1]	0.7	.25	1.5	0.22					
	TumanII[1]	0.4	0.08	2.0	0.05					
1972	ATC[2]	0.88-0.35	0.17-0.11	2.0-5.0	0.11-0.28		0.1	0.16		0.2
	Cleo[3]	0.9	0.18	2.0	0.12		0.4			0.4
	DII[2]	0.60	0.10	0.95	0.21					
	JFT-2[8]	0.9	0.25	1.8	0.23		1.5	1.0	0.3	0.2

	T-12[1]	0.36	0.08	1.0	0.03					
	TO-1[1]	0.6	0.13	1.5	0.07					
1973	AlcatorA[2]	0.54	0.1	10.0	0.31			0.1	0.1	
	Pulsator[2]	0.7	0.12	2.7	0.09					
	TFR400[4]	0.98	0.20	6.0	0.41		0.7			
1974	DIVA[8]	0.6	0.10	2.0	0.06	x				
	Petula[4]	0.72	0.16	2.7	0.16				0.5	
	Tosca[3-]	0.3	0.09	0.5	0.02					0.2
1975	DITE[3]	1.17	0.26	2.7	0.26	x	2.4			
	FT[5]	0.83	0.20	10.0	0.80				1.0	
	PLT[2]	1.3	0.40	3.5	0.72		3.0	5.0	1.0	
	T-10[1]	1.5	0.37	4.5	0.68					1.0
	T-11[1]	0.7	0.22	1.5	0.17		0.7			
1976	JIPPT2[8]	0.91	0.17	3.0	0.16		0.1		0.2	
	Microtor[2]	0.3	0.10	2.5	0.14		0.5			

(Cont.)

First Year	Device	R (m)	a (m)	B (T)	I (MA)	Divertor	NBI P(MW)	ICRH P(MW)	LHR P(MW)	ECR P(MW)
	TNT-A[8]	0.4	0.10	0.44	0.02					
1977	ISX-A[2]	0.92	0.26	1.8	0.22					
	Macrator[2]	0.9	0.40	0.4	0.12			0.5		
1978	ISX-B[2]	0.93	0.27	1.8	0.24		2.5			0.2
	TFR600[4]	0.98	0.20	6.0	0.41			1.5		0.6
	TUMANIII[1]	0.55	0.15	3.0	0.20					
	Versator[2]	0.4	0.13	1.5	0.11				0.1	0.1
1979	AlcatorC[2]	0.64	0.17	12.0	0.90				4.0	
	PDX[2]	1.4	0.45	2.5	0.60	x	7.0			
1980	ASDEX[6]	1.54	0.40	3.0	0.52	x	4.5	3.0	2.0	
	DIII[2]	1.45	0.45	2.6	0.61	x	7.0			2.0
	TCA[13]	0.61	0.18	1.5	0.17					0.4
1981	LT-4[11]	0.5	0.10	3.0	0.10					
	TEXT[2]	1.0	0.27	2.8	0.34	x				0.6
	T-7[1]	1.22	0.31	3.0	0.39				0.25	

Year	Device									
1982	TFTR[2]	2.4	0.80	5.0	2.2		40.0	11.0		
1983	HT-6B[9]	0.45	0.12	0.75	0.04				0.1	0.1
	JET[7]	3.0	1.25	3.5	7.0	x	20.0	20.0	7.0	
	TEXTOR[6]	1.75	0.46	3.0	0.6		4.0	4.0		
1985	HT-6M[9]	0.65	0.2	1.5	0.15			1.0	0.15	0.2
	JT60[8]	3.0	0.95	4.5	2.3	x	20.0	2.5	7.5	
1986	DIII-D[2]	1.67	0.67	2.1		x	20.0	4.4		2.0
1988	Tore Supra[4]	2.37	0.80	4.5	2.0	ergodic	1.7	12.0	8.0	
	T-15[1]	2.40	0.70	3.6	1.0		1.0			2.0
1989	COMPASS[3]	0.56	0.21	2.1	0.28	x			0.6	2.0
	RTP[12]	0.72	0.16	2.5	0.16					0.9
	ADITYA[14]	0.75	0.25	1.5	0.25			0.2		0.2
1990	FT-U[5]	0.93	0.30	8.0	1.3			2.0	4.0	1.0
1991	ASDEX-U[6]	1.65	0.50	3.9	1.4	x	10.0	6.0		0.5
	JT60-U[8]	3.4	1.1	4.2	5.0	x	40.0	7.0	8.0	

(*Cont.*)

First Year	Device	R (m)	a (m)	B (T)	I (MA)	Divertor	NBI P(MW)	ICRH P(MW)	LHR P(MW)	ECR P(MW)
	START[3]	0.2–0.3	0.15–24	0.6	0.12	x				0.2
1992	TCV[13]	0.88	0.25–0.7	1.4	1.2					4.5
1993	KAIST[10]	0.53	0.14	0.5	0.12					
	AlcatorCM[2]	0.67	0.22	9.0	1.1	x		4.0		
	ET[2]	5.0	1.0	0.25–1.0	0.045		1.0	2.0		
2006	EAST[9]	1.7	0.4	3.5	1.0	x		3.0	3.5	0.5
2009	KSTAR[10]	1.8	0.5	3.5	2.0	x	14.0	6.0	3.0	4.0
	SST-I[14]	1.10	0.2	3.0	0.22	x	0.8	1.5	1.0	0.2
	T-15[1]	2.43	0.42	3.5	1.0	x	9.0		4.0	7.0
2018	ITER	6.2	2.0	5.3	15.0	x	50	40	40	20

Country: [1]USSR, [2]US, [3]UK, [4]France, [5]Italy, [6]Germany, [7]European, [8]Japan, [9]China, [10]Korea, [11]Australia, [12]Netherlands, [13]Switzerland, [14]India

(Nomenclature: R = major radius, a = minor radius, B = toroidal magnetic field, I = plasma current, NBI = neutral beam injection, ICRH = ion cyclotron resonance heating, LHR = lower hybrid resonance heating, ECR = electron cyclotron resonance heating. Units: m = meters, T = Tesla, MA= million amperes, MW = million watts)

Appendix E
Awards to the Author for
the INTOR Workshop

The role of the INTOR Workshop in the development of fusion energy is attested in the following two awards presented to the author by the U.S. Department of Energy for his work in the INTOR Workshop.

U.S. Department of Energy
Certificate of Appreciation
Presented to W. M. Stacey, Jr.
For your dedicated efforts during the International Tokamak Reactor (INTOR) study. Your work has pioneered the process of four-party interactions in fusion research that has opened the way for us to proceed with the International Thermonuclear Experimental Reactor program. Also, your efforts in identifying differences and similarities among the national approached to the next major device in fusion have set the tone for ITER and allowed us to expect that the ITER work will be successful.
Signed James F. Decker
Acting Director Office of Energy Research
November 1988

U.S. Department of Energy
Distinguished Associate Award
Presented to Dr. Weston M. Stacey, Jr.
In recognition of your outstanding technical and managerial contributions over the last ten years to the International Tokamak Reactor (INTOR) design program and your longstanding commitment to the worldwide effort to develop fusion energy.
Signed James D. Watkins
Secretary of Energy
January 1990

Glossary

Adiabatic compression	Method for heating a plasma by compressing it by pulsing a magnetic field
Argonne	Argonne National Laboratory, Chicago, IL, USA
Artsimovich	Lev Artsimovich, Russian inventor of the tokamak concept
ASDEX	Axisymmetric divertor experiment at IPP Garching, FRG
Beta	Ratio of the plasma pressure to the magnetic pressure
Blanket	A component around the toroidal plasma that produces tritium and removes heat
Burns & Roe	Engineering company, USA
Cadarache	ITER construction site in the south of France
CEA	Commissariat a l'energie atomique, France
CEC	Commission of the European Communities, Brussels, Belgium
CEN/SCK	Centre d'etudes nucleaires, Belgium
CNR	Consiglio Nazionale delle Ricerche, Italy
CRPP	Centre de recherches en physique des plasmas, Switzerland
Culham	UKAEA Culham Laboratory, UK (host of JET)
DAE	Danish Atomic Energy Commission Research Establishment
D+D fusion	Fusion of two deuterium atoms, which requires a higher temperature than D+T fusion

DEMO	The commercial demonstration fusion reactor that will follow ITER
Deuterium	Heavy hydrogen that is part of the fuel for D+T fusion
DIII-D	Tokamak developed in the 1970s at General Atomics, USA
DITE	U.K. tokamak
Divertor	A component that removes escaping particles and heat from the plasma chamber
DoE	U.S. Department of Energy
D+T fusion	Fusion of deuterium and tritium, which requires a lower temperature than D+D fusion
Ebasco	U.S. engineering company
EC	European Community
ECN	Energy Research Foundation, Netherlands
Efremov	D. V. Efremov Scientific Research Institute of Electrophysical Apparatus, Leningrad, USSR
EG&G	U.S engineering company that managed Idaho nuclear facilities
ENEA	Comitato Nazionale per la Ricerca Energia Nucleare, Italy
ERM/KMS	Ecole royale militaire, Laboratoire de physique des plasmas, Brussels, Belgium
EPR	Experimental power reactor
EPRI	Electric Power Research Institute, USA
ETF	Engineering Test Facility, EPR of the U.S. DoE
ETFDC	ETF Design Center (later FEDC), Oak Ridge, TN, USA
ETR	Engineering Test Reactor, formerly FED, USA
EURATOM	European Atomic Energy organization
FED	Fusion Engineering Device (formerly ETF), USA
FEDC	Fusion Engineering Design Center (formerly ETFDC), Oak Ridge, TN, USA
FER	Fusion Energy Research (Facility), Japan
First wall	The first material wall facing the plasma

Fluence	Product of the instantaneous neutron flux and the time of irradiation
FOM	FOM-Institute voor Plasmafysica, Netherlands
FRG	Federal Republic of Germany (1949–1990), which merged with the (East) German Democratic Republic in 1990
Georgia Tech	Georgia Institute of Technology, Atlanta, GA, USA
General Atomics	General Atomics Co., San Diego, CA, USA
Grumman	Northrop Grumman aerospace company, USA
Harwell	UKAEA Harwell Research Establishment, UK
Hitachi	Hitachi Co., Ltd., Japan
HMI	Hahn-Meitner Institut fur Kernforschung GMBH, Berlin, FRG
IAEA	International Atomic Energy Agency of the UN, Vienna, Austria
ICRF	Ion cyclotron resonance frequency
IFRC	International Fusion Research Council of the IAEA
Ignition	The condition when the self-heating of the plasma by fusion just balances the cooling of the plasma by transport and radiation losses
Impurity	An unwanted atom of material in the plasma that was sputtered from the wall
INTOR	International Tokamak Reactor
INTOR Participant	Official delegate invited by IAEA to attend INTOR Workshop
INTOR Party	The countries that participated in INTOR—EC, Japan, USA, USSR
IPP	Max-Planck-Institut für Plasmaphysik, FRG
ISX-B	Impurity Studies Experiment, a tokamak at Oak Ridge, TN, USA
ITER	International Thermonuclear Experimental Reactor,
ITR	Ignition test reactor
JAERI	Japan Atomic Energy Research Institute
JET	Joint European Torus, Oxford, UK

JRC	Joint Research Center, EC
JT60	Japanese Tokamak 60, Naka, Japan
Kawasaki	Kawasaki Heavy Industries, Japan
KfA	Kernforschungsanlage Julich GmbH, Julich, FRG
KfK	Kernforschungszentrum Karlsruhe GmbH, Karlsruhe, FRG
KTH	Royal Institute of Technology, Stockholm, Sweden
Kurchatov	I. V. Kurchatov Institute of Atomic Energy, Moscow, USSR
Livermore	Lawrence Livermore National Laboratory, Livermore, CA, USA
Los Alamos	Los Alamos National Laboratory, Los Alamos, NM, USA
LUXATOM	Societe luxembourgeoise pour l'industrie nucleaire, Luxembourg
MIT	Massachusetts Institute of Technology, Cambridge, MA, USA
Mitsubishi	Mitsubishi Group (Mitsubishi Atomic Power Industries, Mitsubishi Electric Co., and Mitsubishi Heavy Industries), Japan
NET	Next European Torus
Neutral beams	Beams of neutral deuterium atoms injected in the plasma to heat it
NIRA	Nucleare Italiana Reattori Avanzatia SpA, Italy
Oak Ridge	Oak Ridge National Laboratory, Oak Ridge, TN, USA
ORMAK	Oak Ridge Tokamak
OTR	USSR experimental power reactor design
PDX	Princeton Divertor Experiment, USA
Plasma	A gas of ions and electrons at thermonuclear temperatures
PLT	Princeton Large Torus, USA
Princeton	Princeton Plasma Physics Laboratory, Princeton, NJ, USA
R&D	Research and development

Relativistic electron beam	Method for heating the plasma by injecting high-energy electrons
Sandia	Sandia Laboratories at Albuquerque and Livermore, USA
Siemens	Siemens AG
SIN	Schweizerisches Institut fur Nuklearforschung, Switzerland
T-11	Early-generation USSR tokamak
T-15	Large USSR Tokamak-15
TFR	French tokamak
TFTR	Tokamak Fusion Test Reactor, Princeton, NJ, USA
Thermonuclear	Temperatures of 50–100 million degrees Kelvin
TMB	U.S. DoE Technical Management Board formed for the FED
Tokamak	Toroidal plasma confinement concept
Toshiba	Toshiba Corporation, Japan
Transit time magnetic pumping	Method for heating the plasma by varying the magnetic field
Tritium	A rare form of hydrogen that is part of the fuel for D+T fusion
TRW	Thompson Ramo Wooldridge, Inc., USA
UCLA	University of California, Los Angeles, CA, USA
UKAEA	U.K. Atomic Energy Authority
USSR	Union of Soviet Socialist Republics (1922–1991)
Warrington	UKAEA Research Establishment Warrington, UK
Wisconsin	University of Wisconsin, Madison, USA
ZEPHYR	proposed EC fusion ignition experiment

Bibliography of Official INTOR Workshop Publications

IAEA INTOR Reports

The official reports of the INTOR Workshop are:

"International Tokamak Reactor—Zero Phase," STI/PUB/556, IAEA, Vienna, 1980

"International Tokamak Reactor—Phase One," STI/PUB/619 , IAEA, Vienna, 1982

"International Tokamak Reactor—Phase Two A, Part I, STI/PUB/638, Vienna, 1983

"International Tokamak Reactor—Phase Two A, Part II," STI/PUB/714, IAEA, Vienna, 1986

"International Tokamak Reactor—Phase Two A, Part III, Vols. I and II," STI/PUB/795, IAEA, Vienna, 1988

Summary Papers of the INTOR Workshop

The following papers were prepared, reviewed, and authorized by the INTOR Steering Committee as official summaries of the INTOR Workshop. Hundreds of other papers by the members and contributors to the workshop were published both as summaries and as reports of the detailed work prepared for the INTOR Workshop.

"International Tokamak Reactor Zero Phase," *Nuclear Fusion*, 20, 349 (1980).

"International Tokamak Reactor—Phase 1," *Nuclear Fusion*, 22, 135 (1982).

"International Tokamak Reactor—Phase Two A, Part I," *Nuclear Fusion*, 23, 1513 (1983).

"International Tokamak Reactor—Phase Two A, Part 2," *Nuclear Fusion*, 25, 1791 (1985).

"International Tokamak Reactor—Phase Two A, Part III," *Nuclear Fusion*, 28, 711 (1988).

"The INTOR Workshop: A Unique International Collaboration in Fusion," Progress in Nuclear Energy, 22, 119 (1988).

Summary Papers of the U.S. Input to the INTOR Workshop

Each of the Parties (USA, USSR, Japan, EC) prepared voluminous national reports of their input to each phase of the INTOR Workshop. Summaries of the U.S. input were published in the following papers:

"INTOR—A First-Generation Tokamak Experimental Reactor," *Nuclear Engineering Design*, 63, 171 (1981).

"U.S. Conceptual Design Contribution to the INTOR Phase 1 Workshop," *Nuclear Technology/Fusion*, 1, 486 (1981).

"The FED-INTOR Activity," *Nuclear Technology/Fusion*, 4, 202 (1983).

"U.S. Contribution to the Phase 2A, Part 2 International Tokamak Reactor Workshop, 1983–85," *Fusion Technology*, 11, 317 (1987).

"U.S. Contribution to the International Tokamak Reactor Workshop, Phase 2A, Part 3, 1985–87," *Fusion Technology*, 15, 1485 (1989).